THE SINGAPORE BEING LIFTED AT DURBAN

Fr. [*See p.* 170]

Twenty Thousand Miles in a Flying Boat

by
Sir Alan Cobham

The Long Riders' Guild Press
www.thelongridersguild.com
ISBN No: 1-59048-100-3

To the Reader:

The editors and publishers of The Long Riders' Guild Press faced significant technical and financial difficulties in bringing this and the other titles in the Equestrian Travel Classics collection to the light of day.

Though the authors represented in this international series envisioned their stories being shared for generations to come, all too often that was not the case. Sadly, many of the books now being published by The Long Riders' Guild Press were discovered gracing the bookshelves of rare book dealers, adorned with princely prices that placed them out of financial reach of the common reader. The remainder were found lying neglected on the scrap heap of history, their once-proud stories forgotten, their once-glorious covers stained by the toil of time and a host of indifferent previous owners.

However The Long Riders' Guild Press passionately believes that this book, and its literary sisters, remain of global interest and importance. We stand committed, therefore, to bringing our readers the best copy of these classics at the most affordable price. The copy which you now hold may have small blemishes originating from the master text.

We apologize in advance for any defects of this nature.

DEDICATION

OUR flight through and round Africa in a flying-boat marked a vital point in the history of African air transport. By completing the trip successfully we were encouraged to go ahead with our plans, and the support of the various Governments concerned in air-route development in Africa was a direct result.

There was a time before the start of the flight when I thought the trip would not be possible, owing to the huge financial responsibility involved, but, as in the past, Sir Charles Wakefield, as he was then, came to the rescue, and that is why we called it " the Sir Charles Wakefield Flight of Survey round Africa." Lord Wakefield is an ardent Imperialist. He believes that air transport is one of the most effective means of bringing the peoples of the world in closer touch, and that it is going to do more to bring about the ultimate peace of the world than any other modern development.

This book is a simple account of just what happened on our twenty-thousand-mile journey in a flying-boat. My wife and I dedicate it to Lord Wakefield of Hythe, C.B.E., LL.D., as some slight tribute to the great part he has played in British aviation.

A. C.

ROUTE OF THE FLIGHT

FOREWORD

By SIR CHARLES CHEERS WAKEFIELD, Bart.

(Now LORD WAKEFIELD OF HYTHE)

SIR ALAN COBHAM is a portent. He is in fact one of a number, all concerned with aviation, and all tending to prove to the observer that a new day is dawning—the day of air mastery. Sir Alan is doing for Britain in the air a work comparable to that of Drake and Frobisher and their successors, who established British supremacy at sea. In this case it is not supremacy by force of arms that is in question, but none the less it is essential to the development of Imperial air communications that we should be first in establishing practicable air routes in directions where our interests are manifold. Sir Alan is an air pioneer. His services to the Empire have been of great practical value, and this book is a characteristically modest yet detailed account of one of his notable achievements.

This flight of survey round Africa had many remarkable features, not least the part played by Lady Cobham, who accompanied her husband throughout the long journey and took her full share of the duties allotted to the members of the expedition. In so doing she established a record—which is still unbeaten—for air travel by a woman.

Travel impressions gained from the air are still sufficiently novel to afford most attractive reading. Sir Alan's flight is likely to have results of even greater value, and before many years have passed it will be possible for this book to be read by travellers in comfortable air-liners traversing much of the territory first surveyed by Sir Alan in the course of his journey. For plans are already well advanced for a series of air-mail services throughout the African continent, in which the Governments of Great Britain, the Union of South Africa, Egypt, the Sudan, Southern Rhodesia, and a number of the Crown Colonies are jointly concerned as supporters.

I am very glad to contribute this brief foreword to Sir Alan Cobham's account of his journey, which will have more than ordinary interest for all those who hope for the speedy development of air communications within the Empire—and their name is legion.

CONTENTS

ILLUSTRATIONS

ILLUSTRATIONS

TWENTY THOUSAND MILES
IN A FLYING-BOAT

Chapter I

THE FIRST STAGE

WE sat in a motor-car together, having embarked on the first stage of a twenty-thousand-mile flight round Africa. This preliminary journey was a short one, being but a three-minute ride from the Bull Hotel at Rochester to the seaplane base of Messrs Short Bros. on the Medway. The expedition was to be different from its predecessors, for my wife was not merely coming to the slipway to say good-bye, as on previous occasions, but was going with me to share all the experiences, adventures, and thrills of the flight, and to help me in all my difficulties and trials.

The farewell luncheon had been a success, and we were on time, thanks to the wonderful chairmanship of Lord Wakefield, who was now the third passenger in the car. It was he who, with his openhearted, patriotic spirit, believing that our flight could only do good for Britain, had come to the rescue at an awkward time in the promotion of the scheme, and by his generosity had made the flight possible.

I do not believe that I am usually a nervous,

excitable sort of person, and at no time on any of my long flights have I experienced a mad desire to rush away on the next hop; but from the moment we left the hotel in Rochester my one wish was to get into the air as quickly as possible.

The three minutes had passed, and we had reached the slipway. Then came handshakes and farewells exchanged with all sorts of people I had never met before —although they appeared very far from strangers: they were all so very kind—and intermingled with the throng were our dearest friends, saying, as they shook hands, such nice things, to which I believe I replied in the most inane terms.

At last we were in the pinnace, and then I felt the trip had really started. In a few more moments I should be clambering on board the Short Rolls-Royce flying-boat which I was about to command on this long pioneer, reconnaissance flight round Africa over uncharted air routes.

On board the pinnace was Sir Sefton Brancker, the Director of Civil Aviation. Although very cheery, as always, he had, at that moment, one great regret, namely, that he was not coming along with us—for in 1924 I had the honour of piloting him on one of those early flights of survey when we flew to Rangoon and back. As we came alongside the flying-boat he said, " Good-bye! In a few months' time I shall be coming down here again to meet you on your return, saying, ' Hullo, Cobham! Back again? ' "

The crew were already on board, and our engines were ticking over while the machine headed into

14

THE START FROM ROCHESTER 14

Pacific and Atlantic Photos, Ltd.

wind. It was prevented from going forward by a line astern, where a boatman was ready to release us the moment we signalled. We climbed on board, and then came a few last words with Mr Lancaster Parker, the old test pilot for Messrs Short Bros. He had nursed this particular flying-boat from the beginning, and had taken her on her first trials more than a year before, so that when he stepped off into the dinghy we literally 'dropped our pilot.'

Everything on board was in readiness, and in a few seconds I was sitting in the dual seat beside my assistant pilot, Captain Worrall. Our engineer, Mr Green, then came forward to tell us that the engines were warmed up and all was ready to start. Conway, in the bows, threw our mooring-line overboard, and at that moment I signalled to the boatman in the dinghy at our tail. Instantly he released us from our stern moorings, and while Worrall opened the engines wide I held our craft head into wind as she leaped forward.

Quickly we gathered speed. In a few seconds we were on the step of our hull, and in a few more I found myself automatically easing this giant craft, which weighed about ten tons, off the water.

Very soon we were fifty feet in the air, flying by the thronged slipway where our friends were waving their farewells. At that moment I was too busy getting the Singapore out of the valley of the Medway to think about waving back. Out of the corner of my eye, however, I could see that my wife was doing these honours as she stood half out of the centre hatchway in the full blast of the wind from the propellers.

We were off and away, and our next job was to reach Southampton. Although our craft was a flying-boat that landed on water only, we were not going to journey right round the coast of Kent and Sussex in order to reach Southampton Water. Instead of that we were going to fly right up the Thames to Reading, and then fly due south for forty miles until we came to our alighting spot by the river Hamble.

It had been arranged that we should fly over London precisely at 2 P.M., and I was keen that we should be on time. As we flew up the wide waters of the Thames from the Medway to Tilbury I could not help thinking how practical was our method of flying, for beneath us was one long, continuous landing-ground. However, a little later my views changed, especially when we neared the Isle of Dogs and flew over the mass of shipping and small craft that seemed to cover the waters of the river.

Quickly our broad river was dwindling to a stream, and although, as seen from London Bridge, there may appear to be a vast expanse of water on both sides, when we were flying in our big craft with its hundred-foot span at a height of 500 feet the Thames looked more like a little canal running between mountains of buildings and under a tunnel of arches.

Quickly we came upon Tower Bridge, and it seemed but a second later when we had passed over London Bridge; then came Blackfriars, and suddenly I found that if I intended to follow the course of the river it would necessitate my doing some steep turns in order to follow in time all the river's numerous windings.

It may seem a fairly long run by car or tram from Blackfriars Bridge along the Embankment to Westminster, but when one is travelling at ninety miles an hour and cutting the corners it is scarcely a minute's flight. Thus it came about that almost before we realized it we were passing by Westminster, and I had a fleeting glance at Big Ben, whose big hand was showing half a minute past two. Two o'clock was our scheduled time, and I was glad we were not late. It was just over twelve months before that I had arrived at Big Ben to scheduled time at 2.15 P.M. on returning from my flight to Australia and back. On that occasion my flying machine had been a small single-engine seaplane—a D.H. 50, weighing in all a little over two tons, for which I had a 400-h.p. Armstrong-Siddeley Jaguar engine. To-day my craft—the Short all-metal Singapore—was perhaps the largest flying-boat of its kind in the world, with two mighty Rolls-Royce engines, giving 1400 h.p. between them, and my whole outfit weighed nearly ten tons.

We had soon passed over London and had left the friendly landmark of the reservoirs at Hammersmith. I say " friendly " because when flying over the land in a hydroplane any open space of water always looks so to the pilot, just as when flying over vast stretches of open sea in an aeroplane even the smallest and most rocky of islands have often given me great comfort when I have seen them ahead on my course standing out of the watery wastes.

Staines Reservoir has always been a wonderful landmark for aviators, but on this particular flight it appealed to me strongly as being more than that. I realized that it had another use in aviation—namely, as an alighting spot for flying-boats. Windsor Castle was soon passed, and by this time the Thames, as seen from the air, had dwindled down to a mere ribbon of water, and I doubt whether it could have accommodated our craft had we wished it to do so. After Reading we turned due southward for a flight of forty miles over the Hampshire fields to Southampton Water, and were soon circling round the mouth of the Hamble—the little river up which it had been arranged that we should anchor for the night.

Owing to a cross-wind blowing from the southeast I decided that it would be better to land in Southampton Water, and then taxi up the Hamble to our anchorage outside Fairey's Aviation Works. Unfortunately it was low tide, and after I had landed I discovered that the wind was far stronger than we had anticipated, and that it would be much safer to be towed up the narrow fairway than to attempt to taxi up under our own power. While we were circling round the mouth of the Hamble a small yacht driven by a petrol motor suddenly appeared on the scene It was making straight for us in an alarming manner, and when I saw a second craft—namely, a fast motor-launch—in the distance coming down the Hamble I realized that the yacht did not belong to the aviation people who were detailed to look after us, but to a private person who had come out with the excellent

SOME PROPELLER!

intention of helping us. Evidently he could not see that we were moving forward, and I judged that if he continued to come on at the same rate he would never be able to turn out of our path, so I started to shout and to wave him off before it was too late.

Anyone who has had anything to do with motor-boats knows the difficulty of shouting from one craft to another when the engines are running, and as he was unable to hear me he endeavoured to come nearer instead of going away. But worse was to happen, for while he was approaching us head on his mate stood in the bows and prepared to throw a line to Worrall, who was standing in the nose of our flying-boat, while only a few feet behind him our two gigantic propellers were revolving.

If the rope merely touched the propellers there was every likelihood of their flying to pieces. I think that in such moments one should be forgiven for what is said or done, and somehow I believe that my voice penetrated even above the din, and furthermore my language seemed to intimate to the gentleman that he was doing something wrong. At any rate, he went astern just in the nick of time, and a few seconds later got stuck on a mudbank. In the meantime the proper towing-launch had arrived on the scene, but being possibly flustered and in a bit of a hurry he missed the line that we threw to him. It so happened that this launch could turn only to port, and having approached us, and leaving us on the starboard side, he had, owing to the fact that he took rather a big angle to turn, to come right out of the Hamble, go into the middle of Southampton Water, do his port turn, and then come back again.

We were now in the narrow part of the river, and, as most folk know, flying-boats cannot go astern. Thus we were quietly drifting, and getting perilously near some large posts. However, by this time our friend in the motor-yacht had got off the mud and was preparing another onslaught. Seeing that the motor-boat was coming back to our aid he tried to get there first, for there seemed to be a competition for the privilege of towing us in. We threw a line to the motor-boat, which somehow he failed to catch, and again he had to go for a trip out to Southampton Water before he could turn round again.

In the meantime the mate on the yacht had put off in a skiff, carrying a line from his craft to us. He successfully reached Worrall in the nose of the flying-boat, and by this time, being only too glad to get anyone to take us in tow, we made fast. Our good friend then began to haul himself back in his skiff to the yacht by pulling hand over hand along the towing-line. This line began to tighten and was lifted out of the water, whereupon he stood up in the skiff in order to reach it, leaving his oars trailing in the rowlocks.

Just at that moment the skipper of the yacht, seeing the motor-boat coming back to take charge of affairs, set off at full speed. There was a sudden jerk on the towing-line, which caught under the chin our friend who was standing in the skiff and threw him on his back in the bottom of his boat. His oars were jerked out of the rowlocks and went floating away, and the last we saw was the skiff drifting out into Southampton Water on the tide, while he was standing up in the skiff once more, shouting and gesticulating, being oarless and helpless.

In the meantime the motor-boat had again come on the scene, and the skipper in the yacht seemed to try to head him off, despite the fact that he had already got us in tow, with the result that they both collided, the yacht managing to carry away the motor-boat's glass wind-screen.

Somehow the towing-line got cut, and broke away. The confusion and language for a few moments were terrible, and in the middle of it all I managed to throw our heaving-line, with its weighted end, right into the cockpit of the motor-boat. This checked the squabble immediately, and in a few seconds they had hauled our own towing-line aboard and made fast. Soon we were quietly under way and being towed gently up the Hamble.

I heard afterward that, while all this trouble had been going on, my wife, who was in the hull of the flying-boat, being a little alarmed at the shouting and language, inquired from Green, our engineer, who had been on many flying-boat cruises, if anything serious was the matter. Green replied that it was quite all right, and that to-day's affair was mild compared with the usual landing and mooring up of a flying-boat. However, I was determined that in future we would have a regular routine of procedure and pick up our own moorings. I detailed off each one of the crew to do a particular job when we landed, and with Bonnett and Conway in charge of the drogues, which they dropped astern when I signalled to them by shaking my fist, with Green standing by to restart the engines in case of emergency, and Worrall right forward in the front cockpit with pole-hook in hand, I was able to taxi the machine and bring it up head into wind on to our mooring. This Worrall always managed to pick up, whereupon the engines were shut off and we made fast.

These arrangements were so effective that, except at Malta, we were able to land and pick up our moorings without a hitch on every occasion throughout the entire journey round Africa.

Chapter II

BORDEAUX AND MARSEILLES

WE were due to start the next morning for Bordeaux, but there was a howling gale from the south blowing at 35 or 40 miles an hour, and it needed but a simple calculation—that of subtracting 40 miles an hour from 90 miles an hour, our cruising speed—to prove that our forward speed in the air on our way to Bordeaux that day would be about 50 miles an hour, which meant that we could not reach Bordeaux, nearly 500 miles away, before dark, besides the unpleasantness of a rough journey in such a gale.

I remember wondering at the time if a similar gale would be blowing on our return homeward, because the delights of having a 40-mile-an-hour gale on our tail, and increasing our 90-mile-an-hour cruising speed to 130 miles an hour, would be well worth the bumpy passage we might have to endure.

We were all a little disappointed at not getting away immediately, but I took advantage of this spare day at Warsash on the Hamble to get my crew together. In peace and quietness, after the rush of the start of the flight, I was thus able to organize the stowage of the baggage and equipment in the hull, and make everything on our flying-boat shipshape. Each one of the crew had his own bay in the hull where he kept his baggage, lifebelt, helmets, and all his personal effects. Then all the lockers were tabulated for the spares of the engine and craft, and a proper place was found for every bit of our equipment.

Apart from the usual spares, we carried a spare propeller, which was stowed in the tail of the hull, and then as part of our equipment we had a collapsible canoe. I also carried two guns on board, and we had all Bonnett's cinematographic gear of tripod, cameras, stocks of film, etc., helping to bring up our load. Then, of course, there was all our sea-tackle, such as heaving-lines, mooring-ropes, canvas drogues and lines, bilge-pumps, and we were very proud of our anchor, which weighed only seventy pounds, but to my mind was equivalent to, and just as efficient as, the ordinary seagoing anchor of five times its weight.

We had to wait another day at Southampton, owing to thick fog in the Channel, for after taking off with a visibility of a few hundred yards, and passing over Calshot, which we soon lost sight of, we were unable to find the Isle of Wight. I considered that this was a bit too bad for us to continue, and so we retraced our original course back to Hamble, where we waited another day.

On the following morning we got off well, and although it was a grey November day we had a clear flight over the Needles, and headed on our compass-course for Guernsey. Worrall and I had not known each other until a few weeks before, and this was really our second flight together. Before starting on the trip we had determined to make good compass-courses, and although we could work only on a dead reckoning, we meant to allow for wind-drift by the simple methods we had both learned in the Service during the War. And so, having lost sight of the Needles, we pushed out into the blue, and kept to the compass - course which we had worked out. Naturally it was very gratifying when Guernsey turned up directly ahead of us, and Worrall and I, as we sat side by side, exchanged winks of satisfaction.

When passing over an industrial centre one can distinctly define the smells of the different factories, even, I believe, when those on the ground cannot smell them, because the odours rise into the atmosphere. Thus it came about that when we passed over the Channel Islands we could scent the faint aroma of flowers in the air. I have had similar experiences when flying over the deserts of Northern Africa. I have come suddenly on a river-bed which has been full of herbage, and the whole of the atmosphere at 2000 feet has been scent-laden.

Soon after leaving Guernsey we saw the coastline of Brittany, which we followed all the way round, *via* Brest, to Bordeaux, encountering the usual mixture of North-west European weather—namely, rain, fog, then later wind, with here and there a gleam of sunshine—until at last the mouth of the Gironde was passed. Then we followed what is perhaps one of the straightest pieces of coastline in the whole world, stretching from the Gironde to Biarritz, until we were parallel with the Hourtin Lake, which is a few miles inland.

25

Here we turned over the forest and landed at the French seaplane base at Hourtin.

I can think of nothing more fascinating and exhilarating than the thrill of landing a flying-boat on the rippling waters of large inland lakes, and taxi-ing and manœuvring a big craft like our flying-boat under such conditions gives one pleasure akin to the sport of yachting.

At Hourtin one could land on the outer lake, then taxi into the inner bay, where the anchorages are placed, and I shall always have vivid memories of the beauty of the place, surrounded by sand-dunes which were completely covered with dense pine forest.

It was here, in a moment of enthusiasm, while climbing about our hull, that I managed to slip into the water, but was luckily saved from a severe ducking by the readiness of my wife, who threw me a rope in the nick of time. I managed to grab this so that I only got one leg in the water, and the stiffness of my canvas overalls prevented me from getting wet.

We were quickly taken care of by our French friends, and after the refuelling and general routine had been done we all went into the village of Hourtin for the night. We stayed at a little village hospice, or country inn, the kind that one meets in no other part of the world but France, where at a moment's notice one can get a good bed and a dinner equal to that provided at even the best restaurant, with a bottle of wine that would do credit to the finest hotel cellar.

However, we were going to be more fortunate than the ordinary traveller, for by telephone we had given them three or four hours' notice of our arrival, and accordingly we sat down to a wonderful seven-course dinner, at which we drank the choicest wines of Bordeaux. We were a merry party. To begin with there was my wife, whose knowledge of French is a thousand times better than mine, and who therefore acted as interpreter. Then there was Worrall, my assistant pilot, who was going to take charge of affairs when the business of the expedition called me elsewhere. Green, the Rolls-Royce engineer, with Conway, his assistant, and, last but not least, Bonnett, our cinematographer, who had come along to make a pictorial record of our flight by arrangement with Gaumont, completed the party. It was a unique opportunity for us all to get to know one another, for we had a long and important journey before us, and I knew that 50 per cent. of the success of the trip would depend on the happy co-operation of the members of the expedition.

When we retired to our rooms that night we all discovered the old-fashioned type of French bed, with a mattress about three feet thick. The whole bed was so high that one needed a chair to climb into it, the main covering being an eiderdown also about three feet in thickness.

These things were rather novel to some of the crew, especially to young Conway, who had just left Derby, his native town, for the first time in his life. As there were no chairs in the bedrooms, the crew had, on turning-in, to go to one another's rooms and give each other a leg up into bed. Conway, being the youngest, and presumably the most agile, had to help the last man up, so that when he arrived in his own room he had, I am told, to climb the bedpost.

We were all up early next morning, and got off well from the Hourtin Lake, and after flying due east over the land to the river Gironde we started more or less to follow its course, and, later, that of the Garonne, on our route to Marseilles, our next stop. It was a bright, sunshiny morning, and the broad river glistened like a ribbon of silver, winding its way ahead of us through the vineyards of Southern France.

When I originally planned the route for our flying-boat cruise round Africa many folk showed surprise when they saw that my course lay in several instances over vast stretches of land, for they seemed to think it all wrong that a flying-boat, capable of landing only on water, should fly over the land. I quickly reassured them by pointing out that on my previous long flights there had been no undue criticism about the routes that I had taken with my aeroplane when there had been long passages over the water, and that, from my own point of view, I should feel much happier flying over the land with a flying-boat than I should flying over vast ocean stretches out of sight of land in an aeroplane. So that with these thoughts in our mind there was nothing really extraordinary when, on nearing Toulouse, we left behind the landing-refuge of the Garonne, and passed over the hundred-mile stretch of land to the shores of the Mediterranean in the region of Narbonne. As I have said, it was a beautiful sunshiny morning. The good visibility made it ideal for aerial photography, and I was full of hopes that Bonnett would be able to get some really good pictures of Toulouse, but more especially of the ancient fortress-city of Carcassonne as we passed over it. There had been a gentle easterly breeze at the commencement of the flight, but as the journey went on this increased in force. The sky became overcast and quite ruined all chances of our aerial photography for that day.

Soon after passing Toulouse I could see that our progress was being hindered considerably, and that we should be long overdue by the time we reached Marignane, on the Berre Lake, the air-port for Marseilles. I knew that Green, our engineer, had not completely filled our tanks for this journey, as it was not necessary. He had taken on board enough fuel for this short hop, with an ample margin of reserve. Before leaving England we had had fitted the latest type of petrol-gauge, so when I told Green that we should possibly be an hour and a half overdue he casually walked over to the gauges, pressed the button, and discovered that the tanks registered but very few gallons, certainly not enough for us to reach our destination. He seemed terribly worried, and repeatedly pressed the button to make the gauges work, but each time the hands failed to register more than a miserable twenty gallons, which as far as we were concerned meant that we had practically nothing, considering that our two 700-h.p. Condors consumed about sixty gallons an hour between them.

Inside the spacious hull fevered calculations with pencil and paper were being made of the fuel consumption per hour of the engines, the mileage to be covered, and the estimated quantity of fuel in the tanks. Every one was tapping the gauges and pulling odd levers, but the stubborn fact remained that, according to the petrol-gauges, there was less than twenty gallons in the tanks. Some suggested that there must be a leak in the petrol-tank; others that the engines must have eaten the fuel instead of consuming it naturally; still others, with long faces, wanted to know how the hull would crumple up when we ran out of petrol and had to alight on Mother Earth.

Worrall and I, as we sat in the pilots' seats, were anxiously searching the horizon for the first sight of sea or water on which we could alight. At last a thin streak of silver could be seen in the distance, and we knew it was the Mediterranean. By this time the gauges were showing hardly any petrol at all, and to say that the feelings of all on board were a little strained is to put the situation very mildly. When we were eventually within gliding distance of the sea, although still a long way from Marseilles, our destination, we were all most jubilant, for the chief danger was over. We crept along the coastline, ready at any moment to land on the water beneath us.

GREEN GOES UP TO MEASURE THE PETROL AT MARSEILLES

30

Every moment we expected that our engines would splutter and we should be forced to land on the sea. The minutes dragged by, and it seemed that we should never reach our destination, and I was only too thankful when at last we got to the lakes which would have made much safer landing than the open Mediterranean.

Our gauges by this time had ceased to function, and we were fearing the worst at any moment. I left Worrall at the wheel and went down into the cabin, and was able to cheer them up by telling them that our destination was in sight. After the last hour's worry it was a tremendous relief to be over the Berre Lake, and, although the air-port was twenty miles away, I seemed not to care two hoots whether she petered out or not. At last we arrived over Marignane, that wonderful flying-boat port and aerodrome combined, and after circling once landed head into wind.

I had seen from the air the buoy allocated to us, recognizing it from the instructions that had been sent ahead, so we had no difficulty in getting moored up. Immediately we had made fast and the engines were shut off I looked for Green to ask him to see just how much petrol we had in the tanks, but he had already got the ladder out, and was mounting up to the top wing, where our petrol-tanks were installed, to ascertain with the dip-stick the true state of affairs. After a few moments he looked down at us with a smiling face, saying that we had about two hundred gallons in the tanks—enough for nearly another four hours' flying. All our worrying and scare had been

C

unnecessary, and we made a note that in future we would keep a strict record of exactly how much fuel we had in the tanks before starting on the flight, calculating our tankage by simple arithmetic on the consumption per hour.

Chapter III

THE MEDITERRANEAN

OUR next day's flight was to take us from Marignane to Ajaccio, in the island of Corsica, and the trip turned out to be a typical example of how changeful weather conditions can be in the Mediterranean in the space of just a few hundred miles.

Ajaccio is only 220 miles from Marseilles, and we expected to be there in well under three hours, even allowing for a head wind. We took off from the lake in a dead calm, and, flying direct over the rocky range of hills that separates it from the Mediterranean, made for the bay in front of Marseilles.

The air was calm, and there was little or no wind, although the sky was overcast. Presently we reached the rocky headland, which we flew round, and then turned eastward. Within five minutes a violent gale was blowing and the machine was being thrown about badly. A little later we were in a tempestuous rainstorm, and both Worrall and I had to take a firm hand at our separate dual controls to keep the machine on an even keel.

Our flying-boat had two pilots' seats and dual control. That is to say, there were two sets of controls joined together, so that it was not necessary for both of us to be in the cockpit at once, although on this

33

occasion both our energies were required. Suddenly there was a violent down-current from the towering cliffs away on our port, and in an instant our left wing dropped to an angle of over 45 degrees. Both Worrall and I were struggling with our aileron control and exerting all our strength to pull the wheel over and lift the wing up again. Although only seconds it seemed like an age before the controls answered; and during this time we were falling sideways. Suddenly the port wing lurched upward, and we were on an even keel once more. Worrall and I gave each other a look which meant that we did not want much more of that sort of thing. Curiously enough, although we travelled for another 20,000 miles, around the entire African continent, we never again experienced weather that worried us from the point of view of ' bumps.' This goes to prove that some of the bumpiest atmospheres in the whole world, for flying, can be met with on the Mediterranean coastlines.

After the storm we went under very low cloud, and then met rain once more. This was later followed by fog, and under trying conditions we held on our course, the wind having changed to all points of the compass since we had started. I estimated that we ought to be sighting Corsica very soon, but owing to the fact that the visibility was so poor I knew that we should not see the rocky shores until we were almost on top of them.

Suddenly the weather cleared ahead, the horizon became brighter, and in ten minutes we had flown into perfect sunshine with a cloudless sky. The

BONNETT TAKES A PICTURE OF WORRALL AND SIR ALAN FROM THE FRONT COCKPIT

34

visibility improved rapidly, and presently, forty miles away, we could see the wonderful mountain outline of the island of Corsica.

But this was not the only change. The weather but twenty minutes before had been bitterly cold, and in less than half an hour the atmospheric temperature had jumped over 30 degrees. Then with interest we watched our oil and water temperatures rise, but this did not worry us, for we had tropical radiators and were equipped for the heat of Africa. We knew that what we were experiencing then would be cold weather to what we were going to get later on.

I always think of Corsica as one of the most beautiful islands in the world, and although mountain scenery may have its charm when seen from the ground, to my mind one gets a far more wonderful conception of a mountainous country from aircraft. The reason for this is quite simple: in a mountainous country one always endeavours to climb the highest peak in order to get the best view, and flying over such country is just like being on a floating mountain peak above everything, with an ever-changing aspect.

As we neared the island we made direct for Ajaccio, which at first is hidden from view, as it nestles on the hill slopes behind a headland in the top corner of the bay. Ajaccio, like all Mediterranean coastline towns, is a picturesque sight from the air, but being completely surrounded by high mountains, and open to the sea only at the south-west, it would be a very difficult spot to fly over if the wind were blowing in any other direction than from the south-west. With the wind coming off the land over those high mountain ranges that lie to the east and north the air would be full of violent down-currents and bumps of every description.

When we arrived a stiff wind was blowing off the sea. This had evidently just sprung up, for although the waters beneath us were a mass of white-crested waves, only a short chop was running, no swell having got up. I glided in and landed in front of the harbour, and as we skimmed over the wave-crests one could hear the clatter on the hull as each foam-topped wave was bridged by our keel. We soon came to rest, and then discovered that the sea was far rougher than we had thought, for the waves breaking on our nose occasionally sent spray right over the cockpit. However, we had no difficulty in turning the craft and picking up our own mooring, although the fun started when the skiff tried to get alongside to take us ashore.

We discovered that everybody seemed to think that because the Short all-metal flying-boat was made of metal it could be treated like a fishing-smack or Thames tug. They did not realize that the skin of our hull was less than a sixteenth of an inch thick. Furthermore, many of them seemed to imagine that the trailing edge of our lower wing, which was only about two feet from the water, could be treated as a buffer against which the bows of their boat could be allowed to bump. Thus it came about that very often our first words of greeting on arrival at a new port to the approaching skiff were shouts and orders for them to keep off.

At Ajaccio the sea was so rough that everybody had, more or less, to jump from the flying-boat as the skiff rose on a wave, and Green, having got aboard the skiff, was unwisely holding on to the gunwale against which our fenders were rubbing, as he waited for the last man to jump. Just at that moment an extra big wave came up, and drove the skiff so heavily against the side of the boat that the fender was squashed flat. Consequently Green's fingers took the weight of the skiff, and were badly crushed between the gunwale and the hull. It shook him severely, but luckily his wounds had healed in a week.

It was at Ajaccio that we met Major Routley, the British Consul, who is not only respected as such in Corsica, but is also one of the leading personalities of Ajaccio. Under his capable wing we were shown round Ajaccio. Although we were ready to continue our journey to Malta the following morning, the adverse weather reports and warnings given to us by the French air-line personnel there deterred us. South of Corsica and along the route that we should have to take a gale was raging, and in view of the lack of harbours we felt that it would be unwise to continue that day.

As the morning progressed the gale got worse, and by the afternoon an enormous swell was rolling into the harbour. Even in the inner cambers the water was rising and flooding over the quayside. They informed us that it was the worst day they had had for twelve months, and that as a rule the seas were always calm at Ajaccio. I had heard that tale many times before, and guessed that I was going to have my usual luck, for I have noticed in my different long flights that I always seem to experience exceptional weather conditions. The inhabitants invariably inform me that they have never had anything of the kind before, or that it is the worst weather experienced for years, or something of that sort.

Before starting I had a cable from Egypt to warn me, as I was landing all the way up the Nile, that the river was lower than it had been for twenty-nine years. I can remember on the India flight that the snow fell in Rumania a month earlier than usual, and, apart from five feet of snow on the aerodrome, it caused unprecedented fogs in the valley to the north down which we had to fly.

In the winter season of 1925–26 I made my first flight to the Cape and back in an aeroplane, passing through all countries south of the equator in the rainy season, and I have since discovered that it is referred to as the deluge year, and the worst on record for heavy rainfall.

On my flight to Australia and back, later in 1926, I flew through the heart of the monsoon in Burma and Siam, and it seemed that at all my stopping-places I always arrived in what the inhabitants informed me was the worst storm they had had that year.

And so on my flight of survey round Africa I was fully prepared to meet the usual 'extraordinary conditions' at all points. I have come to the conclusion that any report I make on the flying conditions of a country, as a result of one of my reconnaissance flights, will be a sound report, and not in the least optimistic, for if there are any exceptional conditions to be met with I am sure to encounter them.

During the afternoon at Ajaccio we saw enough of this lovely place to stimulate our imagination and to make us want to return for a longer stay, so we made a mental note that it should be one of our future holiday resorts.

During the afternoon I cabled to Malta for a weather forecast for the following day, our route being from Corsica *via* Sardinia to Sicily, and then across to Malta. By midnight I had a reply, but it was in code and I could not read it, so I had to wait until the morning, when the British Consul, with the aid of his books, and tremendous labour, deciphered it for us. The report was very favourable, and it appeared that along the whole route the winds were going to be light, variable, but mainly from the southwest, which naturally would help us as our course was mostly easterly.

Having read the deciphered report, we all journeyed down to the quayside in good spirits, with the prospects of an enjoyable trip to Malta, a 540-mile journey, before us. There was a gentle breeze off the land, which meant that I should have to go right out of the harbour a long way into the open bay, so that when I pointed my nose inland, head on to the breeze, I should be able to get off and turn before I reached the mountainous slopes that surrounded the bay on all sides. All went well, and we had a good take-off, turning easily. Soon we were flying along the western coast of Corsica with a gentle following wind.

Unfortunately the sky was overcast, which rather hindered us from taking pictures of the marvellous scenic beauty of this coastline, especially the medieval town of Bonifacio, built on the very edge of the cliff, which rose for hundreds of feet sheer out of the waters of the Mediterranean.

Soon after this we turned eastward through the Straits of Bonifacio, passing between the islands of Corsica and Sardinia. Then we followed the east coast of Sardinia southward, continuing our flight with a gentle northerly breeze behind us under ideal flying conditions, until we came to the corner of the island, at Cape Carbonara, when we took up our south-easterly course for the island of Sicily.

As we turned out over the open Mediterranean we quickly lost sight of land, and contented ourselves with keeping a strict compass-course that should bring us out to the western corner of the triangular island of Sicily.

We had not been flying very long before we noticed that white horses were springing up on the wave-crests. This meant that the wind was increasing in strength, but to our great disappointment we could see by the line of the wave-crests that the wind had got round to the north-east. As we were going south-east this was now on our beam and not helping us so much, but worse was to follow, for in half an hour the wind had got right round into the east, and a little later was blowing a gale from the south-east, which meant that we were heading straight into it.

The visibility was poor, but owing to the fact that we were over the open sea, miles away from land, we were having a fairly smooth passage, despite the gale we were heading into. We were flying very low in order to avoid the maximum force of the gale, for the wind always increases in force with altitude. The sea, a few feet beneath us, looked very angry; in fact, it was a mass of giant foam-crested waves, and on one occasion when Worrall and I had been contemplating together the angry waters beneath us he shouted in my ear, " Lord bless our Rolls-Royce Condors! "

The wind increased in force, and it seemed that we should never sight Sicily. I went below on more than one occasion, and at our chart-desk checked up our drift in order to be sure that we were making a good course. According to the time flown we ought to have sighted Sicily long ago, and, being overdue, we kept looking ahead, and I suppose the wish was father to the thought, for we fell into the old error so often made in such circumstances — namely, we imagined that we could see land when it did not exist.

We were an hour and a half overdue when at last I spotted what I thought was land on the horizon, but I did not mention the fact until I had watched it for a considerable time and could discern a rocky out-line going down into the sea. Then I knew that this must be land and not a cloud. The moment I was sure about this I put my head down into the cabin and cheered them up by shouting, " Land ahead! "

What we could see turned out to be the mountain rock-island of Marittimo, off the western extremity of Sicily. It seemed as though we should never reach land, for as the whole coastline of Sicily came into view the minutes dragged on, and yet we appeared to get no nearer. When at last we reached its shores near the town of Marsala I went below to calculate exactly what speed we were making, for I feared that, at the present rate of procedure, although we might have enough fuel on board, we might not reach Malta before dark.

However, after working it out I reckoned that we could just about make our destination, and decided to carry on, for there was no refuge where we could land our flying-boat along the southern coast of Sicily. If I did not intend to make Malta, therefore, I should have to turn back there and then to one of the sheltered ports on the leeward side of Sicily.

It was slow work ploughing our way through a head wind of from thirty-five to forty miles an hour all down the southern coastline of Sicily. The day was dreary, dull, overcast, and misty, and the daylight was beginning to fade when at last we sighted Cape Scaramia, the point at which I meant to alter our course and strike out southward over the sea for Malta.

By this time we were all a little anxious lest we should not sight the island before darkness set in, for unfortunately daylight was fading earlier than usual, owing to the dark clouds in the west and the general bad visibility.

IN A FLYING-BOAT

It was our custom to nurse our engines for the long expedition ahead of us by running them well throttled back, but now was an occasion which justified using the reserve power we had in hand in order to make time. So, opening our engines out and putting our nose down, flying low in order to get the least amount of head wind, we pushed our giant flying-boat up to 110 miles an hour as we flew only thirty feet above the water.

I went down into the cabin for a few seconds where it was difficult to see, but what made me realize how near was the approaching darkness was when my wife, who had been sitting in the front cockpit, came and told me that she had looked back at the engines and thought there must be something wrong, because the exhaust pipes were on fire. Now, although in the daytime the exhaust pipes in an aero-engine may appear to be just ordinary dull black metal, when one is flying in the darkness at night these same exhaust pipes appear to be a glowing red-hot mass with a sheet of blue flame coming out of the end. In reality they are exactly the same when one is flying in the daytime, only the light of day overpowers the glow, and it is not until evening draws on and the light begins to fade that the glow of the exhaust becomes visible.

We were all glad when we saw the outline of the Island of Gozo, because, despite a good compass-course, it would have been quite possible to miss such small islands owing to the approaching darkness, the bad visibility, and a violent cross-wind.

Chapter IV

STORM AND STRESS AT MALTA

OUR intended landing-place had been the Cala Frana seaplane base at the far end of the island from where we had arrived, but as I knew that I should not have time to make it I determined that we must get down as quickly as possible in the first sheltered cove on the lee shore before darkness set in, because owing to the weather conditions I knew that there would be a 'black-out' in a few minutes.

While Worrall took control and flew along under the rocky cliffs I switched on my light and consulted our chart to see just how far away was St Paul's Bay. There I knew, with the wind in its present direction, we could count upon calm waters.

Quickly we passed from the Island of Gozo across the narrow straits to Malta. Then came Melleha Bay, and a mile or two farther on I knew must be St Paul's Bay. As we rounded a little headland a whole lot of lights came into view, which made me think it must be some seaside resort on the far shore of St Paul's Bay, and so we made straight for these lights.

Darkness was coming on quickly, the water was beginning to look black, and I realized that within a few moments it would be impossible to see it, and so we had to make the bay and the lights ahead within the next few minutes.

44

Suddenly we discovered that the lights were not from a town on the shore, but were coming instead from a ship anchored in the bay, and, as we got nearer, to our surprise we found it was a battleship. We circled once, and without wasting a second we landed alongside, aided in seeing the water by the reflection of the lights on it.

Once we were down we were all relieved to be free for the moment from a rather awkward situation, for had we had to land in the pitch dark we would have run a great risk when trying to land on water that we could not see, and our only hope would have been to have gone to Valletta Harbour, where the waters are more or less illuminated by the numerous lights of the town and shipping at anchor there. But here we should have run the risk of colliding on the water with the numerous small boats and dghaisa-men plying for hire in every direction.

With our engines quietly ticking over we kept our nose into the wind, gradually getting nearer and nearer the huge man-of-war. We could see all hands aboard watching us, and we decided that the best thing to do would be to ask permission to tie up astern, so we fired a red rocket as a signal that we required their help.

A few minutes later a pinnace, which was alongside, was under way and making for us, and as it got nearer we discovered that it was manned by the young officers of the gun-room. We asked them to take us in tow, which they did immediately, and then as we neared the battleship I shouted to a group of officers on board, " We should like to tie up astern for the night. Is it possible? " In the dim light I could not see to whom I was talking, but after a few moments a cheery reply came back, " Certainly, and Captain James sends his compliments and would be pleased if you would come aboard."

By this time it was pitch dark, but the difficulty was suddenly relieved by a big searchlight which was thrown on us while we manœuvred and made fast. The sea was fairly rough, so that it was quite impossible for the pinnace to come alongside. This caused a little delay while a skiff was lowered, during which time we covered up our cockpit, inspected our mooring-tackle, got our baggage out, and made our flying-boat safe for the night.

Eventually my wife and I got into the skiff, and were soon climbing up the companion-way, at the top of which we were greeted by Captain James and Commander Kekewicke of H.M.S. *Queen Elizabeth*. In a few brief words we introduced ourselves and I explained our difficulties, whereupon the Captain suggested that we should all come aboard, kindly extending an invitation to dinner. Accordingly Worrall and the crew followed us.

Baths were the order of the day, and by the time we had all finished and, having made ourselves presentable, had adjourned to the Captain's cabin, we were informed that the Commander-in-Chief at Malta, whose flagship the *Queen Elizabeth* was, had been notified of our arrival and had offered Lady Cobham the hospitality of his quarters for the night.

Of course we had to explain the whole day's adven-

TIED ASTERN OF THE "QUEEN ELIZABETH" IN ST PAUL'S BAY

tures and how it came about that we were late in reaching Malta, for the journey from Ajaccio to Cala Frana seaplane base was a matter of 540 miles and should have taken us about six and a half hours. Instead, owing to the violent head wind, it had taken us eight hours forty minutes, making us two hours overdue.

Just before dinner the Air Officer Commanding R.A.F., Mediterranean, arrived with his Staff Officer, and that evening we had a very jolly dinner. Very rarely does a lady dine on board a battleship at sea in the ward-room, especially when she is the only lady present.

After dinner it was arranged that as the *Queen Elizabeth* had to put to sea at nine o'clock the next morning, for gun practice, we should be up early, cast adrift, and take off from the sheltered waters of St Paul's Bay, flying over to the Royal Air Force seaplane base at Cala Frana.

Before turning in that night the Commander was keen that Lady Cobham should make a tour of the ship. To this she gladly agreed, and from deck to deck she was escorted, being given a hearty reception in every quarter by over a thousand men that helped to make up the ship's crew. However, the gun-room officers seemed to have a prior claim upon us, for they maintained that they had rescued us in the first place. Throughout our stay at Malta we always had the happiest associations with the *Queen Elizabeth*, and memories of the timely hospitality there received will long remain with us.

The next morning the same breeze was still blowing, and, saying good-bye to our new friends, we were towed away by the ship's pinnace toward the shelter of the inner part of St Paul's Bay.

Owing to the rough sea at the mouth of the bay it had been extremely difficult to board the flying-boat. Each member of the crew had jumped for it, with the exception of Lady Cobham and myself, who remained on board the pinnace with the baggage, for I wanted to advise the naval officer in charge on the best methods of towing the flying-boat.

Towing a flying-boat in a rough sea with a swell running is no easy matter, and the real danger lies in turning across swell, for as long as one keeps head on to it the flying-boat can ride it fair and square. However, on this morning, we edged off in a diagonal line in order to get the lee of the headland, and once in the protected waters we turned down wind and soon came to the inner reaches of St Paul's Bay, where we had calmer waters to take off from.

However, there was still a stiff chop running, and not wishing to detain the pinnace longer than necessary, or run the risk of damaging our craft by bringing a big boat alongside while all the baggage was transferred, as there was a landing-stage near by, I decided to send Lady Cobham ashore with all the baggage to make her way to the R.A.F. depot by car.

In the meantime we went near enough alongside the flying-boat to give me a second in which to jump for the Singapore's hull, which I did, and landed safely.

48

IN A FLYING-BOAT

By the time we had started our engines up and were ready to take off my wife was already ashore in a local Ford, speeding up the hill with the baggage piled on top, while the pinnace was half-way back to H.M.S. *Queen Elizabeth*.

We opened out and were in the air in a few seconds, and in a few more were circling round the battleship waving farewell. After this we headed along the coastline, flying over the Grand Harbour of Valletta, and fifteen minutes later we were circling over the Marsa Scirocco Bay, at the south-east corner of the island, in which the R.A.F. seaplane base at Cala Frana is situated.

I now come to one of the most distressing parts of our whole cruise, and before going on with the story of the disasters that followed I want to explain briefly how this flight was brought about, in order that our trials and worries will be better understood.

In 1926 I flew from England to Australia and back on a D.H. 50 seaplane fitted with an Armstrong-Siddeley Jaguar engine, and it had been in connexion with the floats of this machine that I constantly came in contact with Mr Oswald Short. The floats were made of metal, for Oswald Short is undoubtedly the pioneer of all-metal construction in Britain, and at that time he was completing the building of the first all-metal flying-boat to be constructed in England. When the Short Singapore flying-boat was launched it was the largest all-metal flying-boat in the world, and after the success of the Australia flight Mr Short and I discussed the possibilities of a long cruise on the Singapore, because there is nothing like a flight across the world to test out aircraft.

Mr Short was anxious to investigate the possibilities of patrolling coastlines by flying-boat, and, further, as there had been no extensive tests of these large craft at high altitudes in the tropics, we came to the conclusion that there could be no better all-round test for the Singapore flying-boat than the flight that we planned round the African continent.

However, the machine was the property of the Air Ministry, therefore we had to obtain their permission to borrow her for this expedition, to which the Air Council, at the instigation of Sir Samuel Hoare, then Secretary of State for Air, readily agreed, providing we insured the craft against loss or damage.

There were many months of hard work in organizing the expedition and corresponding with officials at the scores of places where we had decided to land. On many parts of the route no aircraft had ever been seen before, and over the greater part of our course we should be the first flying-boat ever to pass. Supplies of fuel and oil had to be laid down at all points, and on top of all this many thousands of pounds sterling had to be found to make this flight of reconnaissance possible.

Thus it came about that, with the enterprise and generosity of Messrs Short Bros. and the Rolls-Royce Company, I was able to begin the organization for the flight. We were nearly ready to start when there was still a big deficit threatening to prevent our venture, but Lord Wakefield stepped into the breach. By his great generosity he made the flight possible, because he believed it would further the development of Empire aviation in Africa.

Somehow the responsibility of this flight was far greater than that attaching to any which I had previously undertaken, the main reason being the great expense of it as compared with the others. The machine was insured for about £25,000, and on top of this was the enormous supplies of fuel and oil that the 1400 h.p. of our two motors would consume. Then, again, there was a crew of six to be kept, insured, and paid, so that the whole affair was a much vaster undertaking than any of my previous expeditions.

When we arrived over the Marsa Scirocco Bay we discovered that it was very much exposed to the south-easterly gale that was blowing, and that a big swell was running. However, in the Marsa Scirocco are three inner bays, and we could see the R.A.F. pinnace, as previously arranged, waiting in St George's Bay to direct us to the smoothest spot for landing. As soon as the pinnace sighted us she fired a white rocket, and so, after circling once, we came down and landed alongside.

A big swell was running at the time, but we made a good landing, and as the pinnace came near they pointed across the open bay in the direction of the Cala Frana seaplane base, and suggested that we should taxi across, a distance of a mile or so. At the time we were being thrown about rather badly, and I considered that it was far too rough to cross the

51

open bay under our own power, as we should run the risk of damaging our propellers with the sea constantly breaking over them.

Therefore I asked them to take us in tow. Shutting our engines off we threw them a line, and a few moments later had made fast, and were being towed at the end of about fifty yards of heavy grass-line.

We were again confronted with the old difficulty of towing the flying-boat across a heavy swell. As long as we could keep her at right angles to the line of rollers our craft would ride them, for pitching would not harm her. The danger came if the flying-boat was turned parallel to the line of rollers, when, owing to her span, the rolling would force her wing-tips under the water, and put violent strains on to the lower planes and wing-tip floats which balanced her laterally on the water.

On our journey across the bay we managed to compromise by making a more or less diagonal course, keeping her nose nearly square on to the waves, but as we progressed across the bay the state of the sea seemed to go from bad to worse, and I began to get a little worried. Unfortunately our diagonal course did not bring us up to the mouth of the Cala Frana camber, the refuge we were making for, but above it. Thus all went well until, on arriving opposite the camber, the pinnace, in order to make it, without any warning suddenly turned at right angles along the swell. This manœuvre brought the hull of our craft parallel with the waves, and we instantly started to roll in a most terrifying manner.

We were rocked from side to side, and our wings were thrown at most alarming angles.

The wind was blowing on our port side, and as the rollers came up from that quarter our port wing would be lifted out of the water and heaved right up in the air; while our starboard wing on the other side would be thrust deep down into the trough, and as the swell suddenly came up under the starboard wing-tip float the strain must have been terrific, as the whole was submerged under the water. After about the third or fourth roll not only our wing-tip float, but the lower plane, was being pushed right under the sea, and, fearing the worst, I screamed to them in the pinnace to get head into wind. However, before we could get back into wind the disaster occurred, for we slid sideways down a big roller, thrusting our starboard wing-tip into the water, and the side-pressure on the whole length of the float was so great that its supports broke and the float was wrenched away.

It was a terrible moment, but further disaster was averted for the time being by the spontaneous action of the crew. Green, Conway, and Bonnett rushed out along the footway of the port lower plane to the wing-tip, thus weighing the wing down and lifting the floatless starboard plane out of the water.

By this time we were head on to the swell once more, but were faced with the problem of getting our craft into the shelter of the camber, which was on our starboard, when not only were we unable to turn in that direction owing to the swell, but furthermore the

53

loss of our starboard float rendered the manœuvre impossible.

Time and again we tried to drift back and so edge our way into the shelter of the camber, at the same time keeping our nose on to the swell. Owing to the direction of the wind being against us we could not succeed, for we dared not run the risk of making the slightest turn with our wing-tip float missing.

Nowhere was there shelter, and I was at my wits' end to know where to go with our crippled craft, when I decided that our only hope was to go right across the bay again to beyond where we had originally landed and get in the sheltered waters under the lee of the Delamara Cliff. Then started a terrible journey over the bay, when we had to make our way slowly, crabwise, holding our floatless wing out of the water with the human load of our crew on the opposite wing.

Our course lay naturally along the swell; therefore, in order to keep our nose head on to it, we had to make a zigzag course, for when we got well into the open, after going diagonally for a little we had to drift back, keeping our craft at right angles to the breakers, and then again we would go forward, all the time edging over to the left.

I shall always look back upon the hour that it took us finally to get to safety as one of the most trying I have ever experienced, for minute by minute the gale seemed to increase in force. I stood in the nose of the hull and endeavoured to shout instructions to the pinnace that was towing us, for it was

RIDING THE SWELL WITH HER STARBOARD WING-TIP FLOAT
GONE

NOTE THE PORT WING WEIGHTED DOWN   54

vital that the towing-line should slacken on occasions when some waves bigger than the rest would lift us with terrible force. At these moments I had to shout to the pinnace to stop pulling, for if any excessive strain were put on the remaining wing-tip float, which was taking the brunt of the load, that might be wrenched away also, and our craft would be lost.

Worrall was now with the rest of the crew helping to balance the wing, and, knowing the seriousness of the situation, I had given orders for the crew to put on their Evans life-jackets.

We had nearly reached the shelter of Delamara Point when we ran into the worst rollers of the whole trip across the bay, and from my position in the nose of the hull I thought at times that the Singapore would never mount the next wave that seemed to tower ten to fifteen feet above me. Somehow each time our nose managed to rise above them, although the real shock came when this colossal weight of water struck our remaining wing-tip float, submerging it completely, and breaking over our lower plane, where the crew clung on to the outer struts.

At such moments I would scream to the pinnace to go slow, for to be dragged and forced through such breakers would certainly have carried our only float away. I must confess that at times I felt sick and shaky with the thought that the next roller would overcome us before we could reach the lee of the cliff.

Minutes that seemed like hours dragged on, and at

55

last the seas seemed to abate. I knew then that we were getting gradually into the lee, and a little later we were creeping through comparatively calm waters, although the swell was still running, and the pinnace dropped anchor as near as possible to the rocky shore, with the cliff towering hundreds of feet above us.

For the time being we were safe and had breathing-space in which to think out our next course of action.

Of course, it meant that our flying-boat would have to come out of the water in order to make the necessary repairs to our float-struts, and to refit the repaired float, which, incidentally, had been rescued by the Air Force men on the far side of the bay.

While we were under the cliff we were quite safe, providing the wind did not back round to the south, when we should again be exposed to the full force of the swell, and our only hope lay in the wind dying down in its present quarter, or changing to the entirely opposite direction, when the waters of the Marsa Scirocco would become calm, and we should be able to tow our craft back in safety into the Air Force camber at Cala Frana and then start repairs.

After mooring the Singapore securely the Air Force willingly placed reliefs and guards at our disposal, and with the pinnace at anchor near by watching, with a motor-boat in attendance, we waited through the day and night for the wind to die down, praying that it would not back round to the south.

On the following morning the easterly gale still raged, and so the flying-boat remained under the shelter of the cliff, riding the small swell quite effectively on the remaining single float, the wing-tip above it being weighted down with two hundred pounds of sandbags in order to keep the floatless wing out of the water.

My one fear was that the wind would change round to the south, and every moment of the next two days I was watching the various flags about the town to see what the wind was doing. It was Sunday, and we had lunched with the Governor at the San Antonio Palace. While we were in the garden the sky suddenly became overcast and dark clouds loomed up. It was evident that we were going to have a rainstorm. The flag was limp at the mast, and I anxiously watched to see which quarter the breeze was coming from, and it was a great relief when I saw the flag straighten out before a westerly wind. This meant a change in the weather conditions, and, with the wind in this quarter, calm waters in the Marsa Scirocco. By late that night the swell had almost gone down in the bay, and Worrall and I discussed the possibilities of getting the flying-boat over early next morning into the shelter of the camber of Cala Frana, in readiness to haul her out of the water on an improvised trolley that our crew had been preparing.

Worrall, not wanting to miss an opportunity, was up before daylight on the Monday morning, and in calm waters, the swell having completely died away, towed the Singapore over to Cala Frana. By 7 A.M. we had her moored up inside the camber.

We went away to breakfast, hoping that the waters

57

would become quite tranquil, for a dead calm was necessary to get the flying-boat on to the cradle and haul her up out of the water, but while we were having breakfast we noticed to our dismay that a wind had sprung up from the south-east again, and by the time we had finished and got down to the base the sea in the bay was already rough. By nine o'clock a swell was running, for the Mediterranean is noted for its rapid changes, especially in the vicinity of Malta. The Commanding Officer informed us that at times the swell came right into the camber, and that twelve months before two motor-boats had been sunk while at anchor there. He advised that we should tow the Singapore away into what he thought would be a more sheltered spot in St George's Bay, one of the smaller bays in the Marsa Scirocco.

Minute by minute the sea was increasing and the swell getting higher, and I feared that it was too rough to risk taking our crippled craft on its single float down wind into the open bay in order to tow her to the other anchorage, which by that time was no better than where she was already moored. If our remaining float gave out our craft would be lost completely, for a flying-boat without its wing-tip floats cannot live long.

All day long the Singapore rode at anchor in the shelter of the seaplane base, although a long swell seemed to get right round the jetty and find its way into the inner waters of the camber.

On the following morning the wind was blowing stronger than ever, and the sea in the Marsa Scirocco

WHAT HAPPENED WHEN THE REMAINING PORT WING-TIP FLOAT
GAVE OUT

THE FIGHT TO BEACH THE SINGAPORE UP THE SLIPWAY

Bay was the worst I had seen there. The waves were breaking right over the jetty, and as the morning wore on two feet of green water could be seen flooding over the top of the little concrete pier that helps to protect the camber. The swell was surging right up on to the tarmac and occasionally breaking over the concrete way in front of the hangars, and although the Singapore was riding well at her moorings, balancing on her remaining float, the chief trouble was that the swell inside the camber was at right angles to the wind. This state of affairs was brought about owing to the buttresses and walls on all sides trapping the heavy seas as they rolled in, and while the Singapore rode head into wind the swell at this point was parallel to her hull and rocked her very severely, putting undue strain on to the remaining float.

We could do nothing, for it was impossible to take her out into the open sea, and out of the question to put her on to the trolley to haul her up on to the slipway. To try to put the new repaired wing-tip float on to the starboard wing by means of a skiff underneath it could not be contemplated, as the sea was rising and falling three or four feet.

We could only hope that the wind would die down and that the port float would hold out.

In the middle of the afternoon Captain Worrall noticed that our only float was more submerged than usual, and, despite the sea that was running, he managed to get on board and climb out on to the wing-tip. He hove off the sandbags, hoping that the

59

float would rise accordingly, but when they were all removed it had no effect, and we feared the float was waterlogged.

He then climbed out on to the opposite wing-tip to see if he could weigh it down, but met with no success. Instead the wing-tip float was gradually sinking deeper, inch by inch, and the lower plane was almost touching the water.

The alarm was given, for it was obvious that something had to be done to save the plane from becoming submerged and possibly overturning, and it was decided that she must be beached on the slipway at all costs. An Air Force sergeant had gone out with Worrall, and was in the nose of the craft endeavouring to get the shackles undone and so to free her from her moorings. Air Force men were running in all directions to get ropes, while others were getting the trolley ready on the slipway to see if it were possible to get the Singapore hull on to it.

By this time the lower plane was touching the water, when suddenly, on the top of a heavy swell, our remaining wing-tip float broke away, and the lower port plane started to become submerged.

The sea in the camber was now so rough that it was impossible to get alongside the flying-boat in a skiff, and a line had to be got from the shore to the Singapore at all costs. Without any warning Bonnett, our cinematographer, plunged in and swam out to her, taking the line with him. He was immediately followed by half a dozen Air Force men, who dived into the swirling waters and managed to climb on

60

board the Singapore, quickly joining Worrall, who was still on the starboard wing-tip. However, the combined weight of six men had no effect in counterbalancing the submerged wing, and they simply hung on, poised twenty feet in the air on the uplifted wing.

Bonnett by this time had got a rope fixed on to the tail, and a line had been taken across the camber to the jetty, so that an endeavour was being made to guide the sinking craft toward the trolley on the slipway, for another rope from the nose of the hull was being held by a gang of men on the tarmac.

The waves were now breaking right over the slipway, and it was as much as twenty men could do to hold the trolley in position as they stood waist-deep in the water. Quickly they got the machine into position, and were just about to put forward a big effort to land the hull on the trolley when a mighty swell surged round them and lifted the flying-boat high in the air and washed the trolley right off the slipway, knocking men overboard in all directions. The sergeant-major was rescued in the nick of time, and narrowly missed being drowned.

The trolley disappeared, and they say it was swept away by the undercurrent, for though it weighed a ton or so it was never found again.

We were now confronted with the problem of landing the flying-boat on the slipway without a trolley, and it was evident that serious damage to the Singapore was unavoidable. However, it was better to salvage and risk damaging her than let her sink before our eyes in the camber.

We quickly decided that we must drag her up the slipway on her hull, hoping that the metal structure would stand the strain, knowing that we were bound to harm the submerged wing as it overhung the slipway edge in shallow waters.

Despite the fact that fully a hundred men were hanging on the ropes fore and aft across the camber they could not hold her in position, and she was being washed to the other side of the camber on to a mass of rocks.

Desperately Worrall and the Air Force men who were with him clung on to the lower plane of the wing that was poised in the air, when every moment it was feared that she might overturn on her back. For a few seconds things looked very black, for the mighty swell surged sideways across the camber, taking the flying-boat with it. The men on the ropes, being powerless to hold out against it, were dragged along the tarmac, at the far end of which there was another slipway.

It seemed that she was bound to go on the rocks at the side of the camber, when at that moment another mighty wave came at right angles to the swell from behind, lifting the flying-boat bodily six feet into the air, and, by sheer good fortune, dropped her right on the slipway that happened to be in front.

By a supreme effort the men on the line to her nose held her fast and prevented the backwash from dragging her down the slipway again, and in those few seconds every man available rushed to this line, so that, when the next wave surged up the slipway, with a mighty pull they dragged the Singapore on the side of her hull, sliding her on her keel and chine, over the concrete to temporary safety.

By this time the submerged wing was overhanging the side of the slipway and was but a few feet from the sea-wall, resting on the rocky shallows. We were thus confronted with the difficulty of lifting the waterlogged wing high enough to slide it over the top of the wall to enable us to drag the hull high and dry on to the tarmac.

The sea was getting worse instead of abating, and now that the hull was high and dry on the sloping slipway the tail was thrust down into the water. We feared that the seas which were breaking over our rudder and tail-plane might twist the hull off the slipway, so we set about dragging her higher up.

Iron girders and planks were brought, and ropes were attached not only to the nose, but also to the struts of the machine, and while three hundred men dragged and tugged, iron girders were ruthlessly thrust under the submerged plane as men endeavoured to lever the wing up.

It was the only course of action open to us, although it was one of the cruellest things I have ever experienced, as we more or less destroyed the lower wing-tip in an endeavour to save the rest of the craft.

However, we got her sufficiently high and dry to be in what we considered a safe position for the time being, with only the wing-tip wrecked, for the hull had stood the gruelling marvellously, having been slid up three-quarters of the concrete slipway with

E

little apparent damage. After tying her down with ropes weighted with stacks of ballast and iron girders, with the wash barely reaching the bottom of our hull, we left her with guards for the night.

At 5.30 the next morning the gale raged worse than ever, and the waves were breaking right over our craft, for the sea was now flooding over the tarmac, and the hangars, which were fifty feet away from the sea-wall, were swamped out.

Wave after wave broke over the slipway, washing up great lumps of rock, and boulders a foot in diameter were being carried about the tarmac like marbles as each new roller flooded over the sea-wall.

It was evident that if this sea continued the Singapore would be washed away if it remained in its present position, and our only hope lay in dragging her right up the slipway well on to the tarmac, risking damaging the hull and wrecking the lower wing as we wrenched it over the sea-wall.

Readily all hands turned out into the rain and gale, and by sheer man-power the Singapore was dragged farther up the slipway until the overhanging wing came flush against the sea-wall, when in desperation, to save the craft from a complete wreck, the wing was levered, smashed, and wrenched until it was clear of the wall, and then the leaning hull, supported by the lower-plane *débris*, was dragged over the concrete tarmac toward the hangars, away from the force of the breakers behind.

Later in the day the wind died down and the sea abated, but it was too late, and I shall always

A LUCKY WAVE LANDS HER ON THE SLIPWAY

AFTER BEING DRAGGED UP ON HER HULL

remember how we, the crew of the Short Rolls-Royce flying-boat, stood, a dejected little party, viewing our wrecked craft.

The "Sir Charles Wakefield Flight of Survey round Africa" at that moment seemed a long way off, but after a while we comforted ourselves that it might have been far worse, for on inspection we discovered that when our metal hull had slid up the concrete slipway the brunt of the weight had been taken on the steel keel and chine, for none of the duralumin plates was punctured, the chief knocks to the hull being caused by the swirling boulders. The upper and lower starboard planes were intact, as was the upper port plane, while on the tail we had only lost an elevator. Summing up the situation, it meant that with a new lower port plane, two new wing-tip floats, new elevators, and a general overhaul of the hull, our flying-boat would be airworthy and seaworthy once more.

There was work to be done, so our dejection did not last very long. Cables were sent off to England, and we set to work immediately, aided by the Air Force personnel, to jack our flying-boat up on to a cradle into position for repairs.

The Singapore was far too large to go into the hangars, so all the work had to be done in the open on the tarmac. In the meantime the staff and workers of Messrs Short Bros., of Rochester, performed miracles, for they cut the material, constructed, and finished the metal wing—an intricate piece of engineering—in fourteen days from the word "Go." It meant twenty-four hours' work a day, and was accomplished only by the united efforts of everybody concerned.

As a result of our first cable Mr Oswald Short had dispatched his chief designer, Mr Gouge, accompanied by Mr Bibby, the Works Manager, direct to Malta for inspection of the craft, after which Mr Gouge returned to England leaving Mr Bibby behind to superintend repairs. The Air Force proceeded quickly with the jacking-up of the machine, and all the crew turned to with a good heart to get the Singapore in condition once more. The hull was repaired where necessary, the wings were made ready to receive the new floats, the tail-plane was overhauled, the *débris* of the wrecked plane removed, and by Christmas a ship arrived from England at Malta with our new 'spares' on board.

During our long stay at Malta, when I was not at work on the machine, I devoted much of my time to the propaganda of aviation on the island, for Malta as an Empire air-port offered enormous possibilities owing to its geographical position. I think I journeyed over every part of the island in quest of favourable aerodrome sites, and investigated every harbour from the point of view of its possibilities for flying-boats and seaplanes, and at the request of Lord Strickland wrote an extensive report on this matter.

There were several wonderful aerodrome sites, and I favoured St Paul's Bay, with the Cala Mistra for the protected anchorage, as the best flying-boat base.

Just before Christmas we had further trouble, for a terrible gale sprang up in the night suddenly and

SIR ALAN AND LADY COBHAM AT WORK WHILE AT ANCHOR

slewed our craft off its cradle, dropping the starboard plane on to its dislodged supports, damaging the rear spar. However, by working hard, a repair was effected, and we were in readiness to install the new wing when it arrived on Boxing Day.

Ill-luck seemed to dog us, however, for when the new wing was unloaded we discovered that the lid of the case was smashed in, something heavy having been dropped on it, with the result that the wing inside was damaged. Fortunately the break was not too serious, and a repair was speedily made.

By early January the Short Rolls-Royce flying-boat was ready to be launched once more, and this job was skilfully managed by the Air Force personnel at Cala Frana, who took charge of the affair. Pounds of vaseline were used to grease the slipway, and slowly the cradle slid down into the water. Being unable to rise, owing to the weight of the girders on it, the cradle was carried on down the slipway under the water, while the Singapore itself floated away.

Then came test flights on which we found our craft perfect in every way, and we made preparations to continue our expedition. Before leaving Malta I had the pleasure of taking up the Commander-in-Chief, Sir Roger Keyes, and his family, also the Air Officer Commanding, Air-Commodore Clark Hall, and members of his staff, and on the recommendation of all concerned we took the Singapore over into what was considered a safe anchorage in St George's Bay for the night, before our departure on the following morning for Benghazi.

A stiff breeze was blowing, so we had put extra lines on to our anchorage in addition to our usual mooring-tackle. After our recent experiences we were determined to do all that we could to ensure safety. I was awakened in the middle of the night by the noise of the gale, and it seemed only a few minutes later when the night porter in the hotel came to me with a message that our flying-boat had broken adrift from its moorings, and had been washed ashore in St George's Bay. I will not dwell upon my feelings in those dark hours, but I rushed down, knocked up a man in a garage, and quickly journeyed to the scene of disaster.

The roadway ran along the low sea-wall at the inner reaches of St George's Bay, and a little later the headlights of the car shone upon the silver wings of the Singapore, which was washed right up on the beach, with her tail looming above the four-foot sea-wall and hanging over the promenade.

It appears that the combined force of the swell and wind had broken the seven-ton cable of our main mooring, and then the other lines had given out one after another, so that the flying-boat had simply drifted back on to the beach. For miles on either side of this point the shore was a mass of rocks right down to the water's edge, with the exception of fifty yards of sandy beach in the far end of the San Georgia, and it was on to this that the boat had luckily drifted.

An airman on the R.A.F. pinnace that was anchored on guard near by said that she was ashore a few seconds after she broke adrift, and at one moment he thought that she would become a wreck as she drifted back past a protruding headland, but as she neared this jutting rock the wing that was about to strike it suddenly lifted out of the water as the craft tilted on to the opposite wing-tip float, the uplifted wing skimming over the rock that would have wrecked the machine.

In the inky blackness before the dawn it was difficult to see the true state of affairs, although the immediate worry was to prevent the hull from being washed broadside on to the breakers. Owing to the fact that the sea-wall was only about three feet high, the upward curve of the tail of our hull rose above it, and the whole of our tail-plane was hanging over the pavement. We gathered what men we could and grouped them so that they stood on the edge of the sea-wall holding on to the struts of the tail-plane in front of them, in an endeavour to prevent the hull from bumping on the stone kerb.

A few more men appeared on the scene, and in the blinding rain and gale we struggled to slew the tail round and to bring the nose head on to the breakers. Then we discovered that the lamp-post was in the way, and it looked as though it would break our elevator. One man feebly cried, " What are we going to do? The lamp-post is in the way." Whereupon I shouted back, " Do? Why, knock it down!" This order was readily responded to, and in a few seconds the iron standard was wrenched from its socket and lay flat.

A few minutes later we had the hull at right angles to the breakers on the beach, and more Air Force men having come to the rescue they were posted, waist-deep in the water, on the wing-tip floats, in an endeavour to hold the craft steady. When the flying-boat had first drifted ashore she had taken a list to starboard, thus for some time the starboard wing-tip float had been thrust deep into the sand, and was unfortunately punctured, owing to the stray rocks and stones that were floating about.

Through the grey dawn we held on, wet and shivering, for the rain poured down in torrents unceasingly, and the wind seemed to blow harder than ever. It was difficult to decide which was the best course of action, for the problem was as follows. We dared not launch her and take her out to her moorings because the starboard wing-tip float was already ruined, and, at the same time, if we waited for the wind to drop, the sea would go down, and she would be left high and dry, and we should be unable to launch her again. Then, too, the sea might get worse and hammer the hull to pieces where she stood. The only course open was to put another float on her and drag her out to sea again.

It so happened that the only float available was one of our old ones that was partly repaired, and when we rushed to the depot and made inquiries we discovered that at least five or six hours' work must be done upon it before it could be fit to put on the wing.

All hands that could possibly work on the float at

one time were put on the job, our only hope being that we could refit the float while the men stood in the breaking surf under the wing, and, furthermore, that the job could be done in time so that we could launch her and get the craft to a safe anchorage before dark.

All through the morning the men took it in reliefs to hold on to the wing-tip floats and tail, for those on the floats were often neck-deep in water as the increasing breakers dashed on the beach all round the machine. The local innkeeper came down to watch events, and I ordered him to fetch half a dozen bottles of the best brandy immediately. When these arrived I issued tots all round to the men, as it was the best thing that could be done to keep out the cold and maintain their spirits.

Just after two o'clock the float was finished, and was rushed down to the beach. Then came a terrible struggle while a dozen men, neck-deep in the water, hung on to the port float, weighing the wing down, and holding the starboard wing in position, balancing the craft on its keel, while Bibby, Green, and other willing helpers staggered about in the swirling sea, endeavouring to fit and bolt the port-float struts to their fittings.

As each increasing sea broke I feared the tail of the hull would be battered in on the sea-wall, as it was going beyond human strength to hold her off. Car seats and padding were placed along the kerb to try to save the shock, and all the time the fight to put on the float was continuing.

It was decided that when all was ready we should be towed off the beach on the end of a long line from a tug at sea, and that we should help in this business by starting up our own engines. In the meantime I had discovered a calm camber near by, which hitherto I had been given to understand was far too shallow for the Singapore. However, being at my wits' end to find shelter, I put off in a skiff and tested it, and found that half-way down the camber I had not less than four or five feet of water, although beyond that it ran shallow. Here was room, shelter, and calm water, so that we could moor our flying-boat in absolute safety. This camber happened to be full of small craft, diving-boards, and rafts, but we had them quickly removed, and it was arranged that after we had been towed out a little into the bay a line should be brought to us from the shore. This would pull us gently sideways into the sheltered camber, while the tug would let us drift back, thus keeping us head on to the wind the whole time.

When at last the float was fixed men remained in the water on either side to balance the craft and to prevent the floats from being damaged on the beach. The rope was attached to the nose of the machine, and with the engines started up I gave the signal for the tug to go ahead. At first nothing happened, for the tug made no impression, and it was not until I opened our own engines full out that we managed to drag the hull out of the sand, where it was fast getting embedded. The blast from our propellers at full throttle blew all the men flat who were hanging on to the tail, and brought the men on the wing-tip floats off their feet in the water. There was a pause while we got temporarily stuck, but we bounded forward again in the nick of time, before our wing-tip float could crash on to the bottom, and were afloat in the roughest sea of all my experience.

Worrall stood in the nose of the hull ready to superintend the ropes for the towing, while I was in the pilot's seat at the engines. All went well for the first 200 yards, when the towing-line from the tug fouled on one of the numerous buoys that obstructed our path. No matter how the men in the tug tried they could not free the rope, and the only thing we could do was to cast ourselves adrift, proceeding past the small rocky headland, and coming up into position with the sheltered camber that we wanted to enter a few hundred yards away on our starboard beam.

In calmer waters I could have turned and taxied straight in, but with the swell that was running and the gale that was blowing this manœuvre was not practicable.

Owing to the strength of the wind and an adverse current it was impossible to drift her back into the camber without outside assistance, and so we waited for the rope that would gently drag us into the camber to be brought to us in a boat from the shore, while we kept our machine head on to the swell with our engines.

While we were being thrown about on the waves I discovered that the current and wind were so strong that we were being drifted in a diagonal fashion to the opposite side of the bay, and it was only by opening my engines well that I could keep in position. Away on our starboard beam we could see four men in a rowing-boat struggling through the heavy seas, endeavouring to tow a line to us, for I discovered afterward that, as usual in such moments, the motor-boat engine refused to start. Minutes dragged on, darkness was coming, and still the four men pulled and pulled, but seemed to get no nearer to us, and I could tell that the weight of the rope they were trailing was becoming so heavy that it was getting beyond their power to row any nearer.

The only thing to be done was for us to go to them. This manœuvre meant drifting back between buoys, then approaching the rowing-boat from behind. Steering the machine while drifting backward in such a gale is almost an impossible thing to do with any accuracy, and one is more or less at the mercy of wind and current. Three times we drifted back, endeavouring to pass between the buoys, and we avoided damaging ourselves on these heavy iron structures only by opening out our engines and going forward again. Our only hope of getting ourselves into a favourable position to drift back successfully and miss the buoys was by taxi-ing across the swell. On one occasion we shipped a big wave which heeled us over, thrusting our wing-tip float under water, and the next moment another wave broke, and I saw two feet of green water go over our lower plane. It made the wing bend, but our metal spars withstood the strain, and once more she rose over the rollers.

IN A FLYING-BOAT

Matters were getting desperate, for the light was fading fast, and in such a sea in the darkness it would be hopeless to find our way into the camber. On the fourth attempt we successfully drifted clear of the buoys, and we were able to taxi over to the boat where the four men were struggling to maintain their position.

As we came near they put forth a mighty effort and managed to row in front of us in an endeavour to get the line to Worrall in the nose, but just at that moment a wave bigger than the rest seemed to throw them out of control, and they drifted down on to us, and it looked as though the boat would be thrown right under our revolving propellers.

It was a terrible moment when we saw the skiff poised five or six feet above us on the top of a wave-crest, for it looked inevitable that in the next second they would be flung on top of us.

Worrall in that second shouted, " Look out for the propellers," whereupon two of the men jumped overboard, and the others lay flat in the bottom of the skiff. In a flash I realized that if the boat was not headed off it would possibly break our propellers, get jammed under the wing, and wreck our wing-tip float, and so, pushing the port-engine throttle full open, I slewed the nose round and rammed the skiff broadside with the bows of our hull. In that instant one of the boys in the skiff made a jump toward Worrall. He missed the bollard, but as he fell into the sea luckily caught hold of the mooring-tackle that was hanging from the nose. The skiff meantime had been rammed clear and had drifted away behind us.

I feared that the men who had jumped overboard would drown, but could do nothing to help them. It was as much as I could manage to keep our craft head on to the breakers.

Worrall was leaning forward endeavouring to pull on board the man who was holding on to the mooring-tackle, but he was so low down he could hardly reach him, and as our nose rose and fell, riding each wave, the poor fellow hanging on in front was carried deep beneath the water.

Worrall is strong, and so must have been the man in the water, for at last, by a superhuman effort, Worrall managed to drag him on to the gunwale just in time, for the poor fellow was on the point of collapse.

While I was endeavouring to keep the machine in position head on to swell, against current and wind, I discovered I was able to achieve the impossible, for, with rudder full over and engines throttled back, I was drifting right into the camber. But a few seconds later I found that my success was due to the fact that we had a line from our bow to the shore, where a party of men were guiding us toward the camber. The plucky man who had jumped from the skiff in an endeavour to reach Worrall had taken the line with him, and had held on to it all the time he was being dragged under the water.

Soon we were out of the main force of the swell, and a few minutes later I had shut off my engines, for we were in our haven of refuge at last, where there were not only tranquil waters, but where we were sheltered from the wind by the houses surrounding this camber.

Very soon it was dark, and although most of us had been out in the continual downpour since three o'clock in the morning, fighting against the elements, we were so happy to have our craft safe at last that with renewed energy we tied up our machine fore and aft to the opposite sides of the camber, and with long lines we made our wing-tips fast to permanent buoys, for, owing to the smallness of the camber, we feared that there was not sufficient room for the Singapore to swing at anchor.

On the following day the sun shone, and it was beautifully calm. After making an inspection we decided it would be best to lift the Singapore out of the water in order to view the bottom of the hull after its bumping on the beach.

Being able to fly the machine meant that we could take off from the Marsa Scirocco and land her in the Grand Harbour. In the naval dockyard there was a crane with a long arm that could lift us on to the quayside. During the day I visited the dockyard authorities, and the Admiral willingly afforded every facility for us to lift our craft when we arrived.

Landing in the Grand Harbour is no easy matter. Firstly, it is a long, narrow waterway with towering cliffs on either side, and invariably occupied by the Mediterranean Fleet. On top of this, difficulty arises from the presence of other shipping, but the real trouble, from the point of view of a flying-boat which wishes to make a landing, comes from the numerous steam-ferries and the hundreds of dghaisa-men with their gondola-like boats plying for hire in every direction.

However, the King's Harbour-master came to our rescue, as also did the Civil Police, and at the appointed hour, having taken off from the Marsa Scirocco, we were soon circling. over the Grand Harbour prior to landing.

One must either land up or down the harbour, as, owing to the high cliffs and lack of width, it is impossible to land across it. There was a gentle breeze down the harbour from the sea. This was most fortunate; and according to plan, in order that all boatmen could get clear, I circled and flew low right down the main fairway between battleships whose masts rose above us on either side. It was arranged that I should do this twice and on the third time land. Having carried out the plan I was gliding in on the third time with apparently a clear fairway ahead of me, getting lower every moment, and was well below the decks of the men-of-war on either side, when I noticed some wretched dghaisa-man rowing right across my fairway. Luckily I had not touched the water and had time to open out and climb away once more. However, on my fourth circle I got down, and even then the same dghaisa nearly fouled my path.

We were soon taken in tow by the R.A.F. pinnace, while a naval pinnace followed up with a line to our stern, and before long we were round the creek by the naval dockyard, where the crane, with a specially made spreader, was in readiness to haul us out of the water.

WORRALL IN THE NOSE OF THE SINGAPORE

There were four lugs fitted permanently to our centre section on the top plane, and very soon we had the shackles fixed and the crane had lifted the Singapore on to the quayside. Quickly she was made secure with trestles and ropes, and we were hard at work inspecting the hull, where we found that two of the plates had been bent and would need repair. However, we were under ideal conditions for working on the hull, and in two or three days the Singapore was ready once more to resume her flight of survey.

She was lowered into the water and towed away to an anchorage in the Grand Harbour in readiness for our flight to Benghazi. Now this flight was one of 450 miles on a south-easterly course over the open sea, out of sight of land the whole way from Malta until we reached our destination at Benghazi. Our only real worry was our landing in Benghazi Harbour, which was open to the swell from the north-west, so that a favourable wind for getting us there was really unfavourable from the point of view of landing in the harbour. Combined with this difficulty there was the problem of getting out of the Grand Harbour, which ran from north-east to south-west. It meant that we must have a wind from either of those directions in order to get off successfully with our full load.

Daily I consulted with the Meteorological Office, who did their utmost to advise us for the best, but when the day came for our start there was a gentle breeze blowing across the harbour, and although we tried to take off under these conditions, across wind, we discovered that not only was it impossible to keep

F   79 our craft straight between the buoys and shipping, but the down currents from the cliffs towering hundreds of feet up on either side prevented the boat from rising.

I had been warned about this by Squadron-Leader Marix, one of the early pioneers of piloting, who happened to be a Staff Officer at Malta, and so we did not make a second attempt, for we knew that we should have to wait until the wind blew either up or down the Grand Harbour, or wait until there was no wind at all.

It is an ill wind that blows nobody any good, for half an hour later, when we were once more moored up to our buoy, the Customs pinnace brought a cable out to us. On opening this I discovered that it was a weather report from the British Consul and our Italian friends at Benghazi, which read as follows: " Northerly gale abating, big sea still running, landing in harbour highly dangerous owing to swell. To land in open sea suicidal." This relieved our disappointment for not getting off that day, and so, accepting the kind hospitality of the King's Harbour-master, whose house was on the quayside, we waited, ready to jump off at the first opportunity.

For three days we waited for favourable weather, and as communication took about twenty-four hours from Benghazi we had to take our chance with the condition of the harbour at that end, intending to fly on to Tobruk, another 250 miles along the coast, if it should prove to be impossible to get down at Benghazi.

IN A FLYING-BOAT

At last, on the morning of January 21, we looked out at dawn to find a dead calm. Very quickly we said good-bye to our friends, and in the harbour-master's pinnace made our way to the Singapore, where our crew had just arrived. In a few moments all was in readiness, and starting up our engines we cast adrift and taxied up to the mouth of the Grand Harbour.

To have taken off in the open sea would have been impossible owing to the long, oily swell, and so, our engines being warm, we turned round and opened out, racing down the whole length of the Grand Harbour. We made a straight line between ships and buoys, and then, gathering flying speed, we rose into the air, off at last from Malta to continue our flight of survey round Africa.

Chapter V

THE LAND OF THE NILE

WE intended taking up our compass-course from Delamara Point, direct over the open sea for Benghazi, and as we passed over the Marsa Scirocco Bay, with the R.A.F. seaplane base at Cala Frana on our right, we all had recollections of days of great trial, and kind memories of the Air Force personnel, who had so willingly stood beside us and helped us to win through. Had it not been for the whole-hearted co-operation of the officers and airmen at that depot our flying-boat would have been lost on more than one occasion.

The sea was calm, there was little or no wind, and while Worrall and I took up our compass-course with careful accuracy, heading out over the open sea, the remainder of the crew were looking out of the rear cockpits, as the Marsa Scirocco got farther away on the horizon behind us.

We were more or less flying into the sun, and the glare was so great that if we looked ahead over the water for any length of time we were unable to see the faces of our compasses, which, in the shaded light of the cockpit, appeared quite black. There were two of them, one on each side, so Worrall and I took it in turns. While one looked ahead and took charge of the wheel, keeping the machine level, laterally and fore and aft, the other kept his head in the shade of the cockpit, with his eyes glued on the compass, keeping the machine on her course with his feet on the rudder-bar.

Hour after hour we carried out this routine, with as great an accuracy as possible, for we wanted to hit Benghazi. In the cabin spirits were high, for the Condor engines were running well, the sea was calm beneath, and we were on our way once more.

Having worked out our approximate time of arrival we began to scan the horizon for the sight of land, and presently we sighted a steamer which was evidently making for the same port as ourselves. Soon we overtook her, and at that moment we could see the clean line of the horizon broken by the rocky hills of the hinterland. Presently white buildings .could be distinguished in the brilliant sunlight, and a town came into view. We knew it must be Benghazi, for there was no other town for hundreds of miles along the coastline. However, we continued to hang on to our compass-course, for we wanted to prove the accuracy of our route, and sure enough it brought us right on to the mouth of the harbour. We could not have gone straighter had we been shot out of a gun. We had hit the African coastline at Benghazi, after 450 miles over the open sea, in exactly five hours' flying.

Soon after we had landed our Italian friends came out to greet us, and gave us a wonderful reception on the harbour steps. Every hour of our brief visit there was filled with various functions.

I shall never forget the *faux pas* that I made while we were guests at the house of his Excellency the Governor. Neither my wife nor myself could speak Italian, and we were very happy when we discovered that the Governor's wife spoke perfect English. At one moment she would be speaking rapidly in Italian to her husband, and would then break off into perfect English as interpreter for our benefit. When I complimented her upon her perfect accent she burst into laughter and said, " Of course it is good. I am English, but am married to an Italian."

That evening we received an urgent cable from the Meteorological Office at Malta telling us that there had sprung up a north-westerly gale, which was moving toward Benghazi. The forecast was that the sea would become enormous, and we were advised to get away as quickly as possible.

Early next morning we were happy to find that the forecasted gale had not yet arrived, but, losing no time, we said farewell, and managed to get off safely from the limited run afforded in the harbour.

I had been along this coastline before, and knew a little about the weather conditions. Fearing that we might get caught out with an unfavourable wind in the open bay at Sollum, where we had originally intended to call, I altered my arrangements so that we landed instead in the magnificent sheltered bay of Tobruk, a lonely Italian outpost on the North African coastline.

Here again we discovered that our Italian friends had made every arrangement possible for our comfort,

PASSING OVER ALEXANDRIA

and that night we slept in the Governor's house high on a hill. After a jolly dinner-party we retired to our rooms, and Worrall having reported that the flying-boat was secure for the night we all turned in.

Soon after midnight we were awakened by a fearful clatter. Windows slammed to, glass was shattered, and we discovered that a terrible gale was blowing. The worries of Malta were still in our mind, and instantly our thoughts turned to our flying-boat anchored in the harbour below. We knew that she was safe from swell and rough sea, but the question was, would the gale break her moorings?

We lay and listened to the wind howling and whistling through the shutters, and being able to stand it no longer I jumped up and said I was going down to the machine. Instantly my wife said, " I am coming with you," and with a few clothes thrown on we were soon groping our way through the windy night toward the harbour.

When we got down to the water's edge we discovered that the wind had abated, for we were in the shelter of the high hill behind. However, we got the guard to find a boatman who, by the light of a hurricane lamp, rowed us out over calm waters to the Singapore, to which we were guided by her riding lights. As we got near we were happy to find that everything was in perfect order and that she was riding peacefully at her moorings. So in high spirits we returned to our beds content with the thought that all was well for the night.

It must have been at about three o'clock in the

morning that we were again awakened by the gale, but this time it was more furious than ever, for when I opened the window I estimated that it was blowing forty or fifty miles an hour. I held a brief consultation with Worrall, who was in the next room. He insisted that it was not necessary for two of us to go down, and said that he would go alone and send up for help if required.

Anxiously we waited for about half an hour while the gale raged, and then Worrall returned, telling us that it was not so fierce on the water, although it was blowing strong, but that the moorings were holding well. I do not know whether we slept much more that night, but we were up with the dawn and soon on our journey once more.

We had a following wind, and made wonderful progress as we flew eastward along the North African coastline. The day was fine, and the colouring effects were marvellous, the deep blue sea contrasting so vividly with the snow-white surf and brightly coloured cliffs of the desert coast. I shall always remember the pleasure I had in showing my wife that wonderful blue lagoon at Marsa Matruh as we flew over. The waters of this sheltered harbour, when seen at an altitude of a few hundred feet, vary in shades from the brightest turquoise to the deepest indigo blue.

By noon we had sighted Alexandria and its great harbour, and as we got nearer we could see battleships at anchor. We had said good-bye to friends on these same ships a week earlier at Malta before they had departed on their manœuvres, and as we approached the harbour we flew low and circled round the *Queen Elizabeth* for the second time on our flight.

We were going to land at Abukir, on the tranquil waters of that sheltered bay that is ten miles beyond Alexandria, and so passing over the town we had soon reached our destination and landed. Here the waters are shallow for miles out into the bay, and we discovered that for some unknown reason the special mooring-buoy put down for us was about a mile from the shore, and as rowing-boats seemed to be the chief means of conveyance it was quite an excursion from the flying-boat to the little wooden pier.

A kindly R.A.F. officer came out to greet us, and after much hard pulling we at last approached the landing-stage, where a little throng was waiting to meet us. We noticed that the gangway was barricaded off on either side, and that our personal friends were behind the rails. The moment we stepped on to the landing-stage we were informed by two officials that we were now in quarantine and must go forthwith to the examining officer's house for inspection. No one was allowed to come near us, and Captain Gladstone, my co-director, who had come all the way from Kenya to meet us, had to wave to me through the railings as we were quickly rushed into a waiting ambulance and driven away a mile or so along the beach.

We had heard some weeks before that there was plague in Egypt, and our friends were horrified at the thought of our going there, but now it seemed that we had the plague and not Egypt. However, the officials had their duty to perform, and the thing to do was to get it over as quickly as possible. The doctor and quarantine officer were very kind, and were as forbearing as possible in dealing with the numerous forms that had to be filled up and the questions answered relating to our personal health and life-history, also that of all our relations to the third and fourth generation.

Then, of course, the Singapore had to be treated like a ship, there being no special forms for aircraft. Marine papers had therefore to be used, and quickly I answered as the official read out question after question on the form—whether or not we had any vermin on board, what was the tonnage of our craft, what was the tonnage of our cargo, etc.

However, even quarantine and customs officials could not defy the Press, and while my wife and I were riding with Worrall in the ambulance a Pressman climbed on the front seat with the driver, and I gave him an interview through the iron grating. Another gentleman of this great fraternity got into the quarantine station, and although not allowed to come near us he interviewed me from the doorway.

Eventually formalities were completed, and we were passed as 'clean.' But we were not out of quarantine until the other members of our crew had been inspected. However, the doctor took pity on us and said he would still keep us in quarantine and under observation by taking us to his house for a late luncheon with his wife and family. To this the quarantine official agreed, providing we promised not to leave the doctor's house until the rest of the crew had passed out. We quite understood the necessity of safeguarding the population of Egypt against the possibility of infection, but presumed that the doctor and his family were somehow immune.

We had left the rest of the crew, Green, Conway, and Bonnett, on the Singapore to await the return of the skiff. Growing tired of this, they had indulged in a bathe, and had gone off in their bathing-costumes for an excursion round the bay in our collapsible canoe. The utmost consternation was expressed on every side, and a search-party eventually found them paddling about on the waters of Abukir Bay without having passed quarantine.

However, they were quickly brought in and taken in the same ambulance up to the quarantine station, and there they went through the same business which we had suffered. This was followed by a long wait for the " All clear," during which time I suppose a conference was held to decide whether we could be released from quarantine, and it was not until the sun was setting that we were at last permitted to drive away to Alexandria in the taxi we had summoned.

That morning we had covered a journey of 450 miles in four and a half hours, and had arrived at Abukir about noon. Yet it took us over six hours to get from the landing-place to our hotel—a matter of five or six miles. Something seemed wrong somewhere, especially when I had the bill for the ambulance ride sent to me. I carefully worked the matter out, and came to the conclusion that at the very utmost the ambulance could have made only four journeys of about a mile and a half each, yet the transport charge for these little rides somehow amounted to £13 10s.

The bill read that at the regulation rate, plus waiting time and attendance for journeys to and from the quarantine station, plus 10 per cent., plus overhead, etc., etc., the total charges came to £13 10s. I had heard that the Air Force had a new form of accountancy, and I happened to be discussing the matter with an Army general a few days later, who did not seem to think it very extraordinary at all, but told me that when he was commanding a unit at Aldershot he had to hire a Ford car from the A.S.C. to do daily work, and when the final bill came through that included wear and tear on car, men's time, overhead charges, and interest on capital, he discovered that it cost him £1500 a year for a daily trip to the station and back. I remarked that £13 10s. for five miles was an even higher rate than that, whereupon the general remarked, " It is evident that the R.A.F. system of accountancy is an improvement upon that of the A.S.C."

The following day Captain Gladstone, of the North Sea-Aerial and General Transport, Ltd., and myself were able to hold a long-delayed conference. His company and mine had recently joined forces on the through - Africa air - route proposition, and it was essential that we should meet for a full consideration of the problems before us.

For years I had been contemplating the inauguration of air routes in Africa, for in 1922 and 1923, when I had made a 12,000-mile tour through Northern Africa from the Atlantic to Egypt and back again, I had been impressed by the possibilities of air-transport in Egypt, and when in 1925–26 I had first flown to the Cape and back I returned with the intention of devoting my energies to air routes in Africa, and formed a company under my own name with that work as one of its main objects.

Thus it came about that in 1927 an amalgamation was made of the African air-route interests of Alan Cobham Aviation, Ltd., with those of the North Sea-Aerial and General Transport, Ltd., on a section of the air route over which they had been running an experimental seaplane service for the Colonial Office.

The new 'holding' company took the name of Cobham-Blackburn Air Lines, with the definite object in view of negotiating and organizing a company that would bring about a through air route from the Mediterranean to South Africa.

Captain Gladstone had been most unfortunate when running the experimental seaplane service for the Colonial Office, and there were still some flights left to be done. As Captain Gladstone's seaplane was out of action an amicable arrangement had been arrived at with the Colonial Office on the advice of the Air Ministry. By this it was agreed that providing we made a flying-boat reconnaissance flight from Alexandria, right up the Nile, through the Sudan to Uganda and Kenya and on to Mwanza, in Tanganyika, and then flew back again all the way to Khartum, returning once more to Mwanza, the old contract that

Gladstone had made on behalf of the North Sea-Aerial and General Transport, Ltd., would be finally cleared up to the satisfaction of the Colonial Office by the new amalgamation of Cobham-Blackburn Air Lines.

Before continuing the flight Gladstone and myself took the first train to Cairo in order to see Lord Lloyd, with whom we had an interview about our proposed air route. I suggested to Gladstone that it would be a good scheme if he flew with us up to Central Africa, and two days later we took off from Abukir, and passed over the Nile delta *en route* for Luxor.

One must really fly in order to get an idea of the reason for Egypt's existence, for an hour after leaving Abukir the majority of the broad delta had been passed over, with its miles of intense cultivation, all of which is accomplished by irrigation from the waters of the Nile.

On either side the desert was closing in on us, and ahead we could see the Pyramids of Gizeh and Cairo.

We did not land at this great metropolis, but, after viewing it from the air, flew over to the Great Pyramids, which we circled twice in an endeavour to get aerial pictures, and then continued our journey up the Nile.

The broad cultivated acres of the delta had gone, and the land of vegetation was now thinned down to a narrow strip of a mile or so on either bank of the Nile. It is easy to realize that without the river Nile there would be no Egypt, for only so far as the waters of this river can be carried by irrigation is

AT REST ON THE NILE AT LUXOR

FLYING LOW: REFLECTIONS IN THE WATERS OF THE NILE

93

there life and cultivation. From a great altitude Egypt after the delta appears to be a broad ribbon of green land winding through the desert, and in the middle of this ribbon runs the Nile, which feeds it.

Egypt is one of the finest flying countries in the world, for on the route down the Nile, generally speaking, it is possible to fly 365 days of the year, day or night.

Soon the Pyramids were far behind us, and as we passed each landmark I could not help thinking how ideal were the conditions, for it was possible at any moment to land on the broad waters of the Nile beneath us should we wish to do so.

The journey from Alexandria to Luxor is one of about sixteen hours or more by train, or a matter of weeks by river steamer, but by air we accomplished this trip in just over five hours, having gained a far better conception of the characteristics of Egypt than one could ever obtain by travelling on the ground.

At Luxor we landed on the river, and picked up our moorings in front of the Winter Palace Hotel. It was all delightfully convenient, and I was more convinced than ever that this was the best way to travel in Egypt.

As neither my wife nor any of the members of the crew had ever visited Egypt before, it would have been a shame to continue the journey without a visit to Karnak, Thebes, and the Valley of the Kings, so a day was spent in visiting these world wonders. On the following morning the journey was continued, and I had the pleasure of pointing out and circling round the temples of Kom Ombo, Esna, and Edfu, until we came to that wonderful aerial picture, the First Cataract at Assuan, with Elephantine Island and the rapids in the foreground, and the great Assuan Dam and Reservoir beyond. So fascinating is this sight that we had to circle to view it again, during which time my cinematographer endeavoured to get pictures on which are seen in juxtaposition this seat of an ancient civilization and one of the greatest engineering accomplishments of modern times.

We read in history that through all the ages there have been years of famine in Egypt when the flood-waters of the Nile have been insufficient to irrigate the land, but with the building of the Assuan Dam, which conserves the waters in the natural rock-gorge reservoir behind it, famine in Egypt has been ended for ever. The Scottish engineer who built the dam remarked that if the ancient Egyptians could build for eternity so could we, for he reckoned the dam would stand for ever.

The reservoir, which is supposed to hold a million million gallons of water, is about ninety miles long, and as we continued our flight up the Nile to Wadi Halfa, on the borders of the Sudan, where we were going to refuel, we passed over many temples of the ancient Egyptians which to-day are half submerged in the waters of the Nile owing to the construction of the dam. At last we came to Abu Simbel, that famous rock temple hewn out of the cliff face, and we could not refrain from coming down low to have a look at it.

IN A FLYING-BOAT

Wadi Halfa, the base of Lord Kitchener's expedition into the Sudan, proved to be an excellent flying-boat base, there being ample room on the Nile in which to manœuvre. The Governor had very kindly made all arrangements for us, and after quickly refuelling and taking a light lunch, that was so thoughtfully prepared under an awning on a launch near by, we took off from the waters of the Nile that sparkled in the midday sun.

When I first asked my wife to accompany me on the tour round Africa I had only one fear for her comfort, and that was that the heat of the Sudan and tropics might be too great for her. As we journeyed day by day farther south I anxiously looked for signs of distress, but to my surprise the hotter the conditions became and the more the temperature rose the better she seemed to like it; in fact, she could stand the heat as well as any member of the crew.

It was a new experience for me to go on a flight of survey and be able to keep pace with all the mass of correspondence and reports that must be kept up to date on such an enterprise. In the past it had meant hours of writing late at night, and stacks of abbreviated notes which had to be rewritten on my return, but now on this flight I was able to dictate to my wife full, detailed reports, while she took them down straight on to the typewriter. Furthermore, very much of this could be done in comfort while we were flying in the air. As is generally known, we had a dual-control cockpit, with two pilots' seats abreast

G

and a narrow gangway between, so that, with a double set of controls, Captain Worrall and I could either fly the machine together or could alternately relieve each other at the wheel while one of us was below in the cabin.

All the way along Captain Gladstone and I were making notes on the possibilities of the Nile at various places, from a flying-boat point of view. After Wadi Halfa there was rather a bad passage over the Second Cataract and beyond, where it would have been impossible to land, and these conditions more or less prevailed until Dongola, but soon the river broadened out again, and an hour before dark we arrived at Karima, where we intended to land for the night.

On coming down from the cool atmospheres of two or three thousand feet we found that the temperature had risen considerably, and that we were in for a hot night. Arrangements had been made for us to sleep on board a small river steamer, but my wife and I found that it was so breathless in the cabins that we got the boys to move the beds on to the top deck. There were no mosquitoes, for they said it was far too hot for them to live in that country, so we lay down in the open, with the stars as a ceiling. I had previously enjoyed similar experiences in the Persian Gulf, and the joys of sleeping in the open in tropical climates where there happen to .be no mosquitoes are one of the happy memories of life.

There was little or no moon, but the stars were so bright that we had no difficulty in seeing our way about, and when just before dawn a bell rang we awoke, very refreshed, in time to see a most glorious sunrise.

At Karima we had rather a bad take-off owing to a cross-wind and the narrowness of the river, and we made several notes about this section of the route. It was our intention to reach Khartum that day, and all went well until we got in the region of Berber, when I noticed that the visibility was beginning to diminish. There were signs of a slight dust-storm ahead, but these rapidly became worse, and in order to follow the Nile we had to come down from our cool altitude and fly low. Very soon it became so thick ahead that I decided to turn about. I then flew back a few miles and landed on the river opposite to Berber.

We dropped anchor a little way from the bank, and here we had the alarming experience of finding ourselves suddenly thrust into a temperature twenty or thirty degrees above that in which we had been flying. The heat on the water was terrific, and as there seemed to be no boat anywhere near I told the crew to get our collapsible canoe out, and I rowed over to the bank, where hundreds of natives were assembled. By the time I was ashore the District Commissioner had arrived on the scene, and, after I had explained to him our reason for landing, he suggested that we should make our craft secure and all come up to his house.

Berber is a huge native town, with the usual sandy, dusty thoroughfares, and therefore walking a mile or so to the Commissioner's house when the temperature was 105 degrees in the shade was not expected of us.

97

Very quickly donkeys were brought, with enormous saddles that seemed to have no girth-band, for the rider had to balance the saddle, as well as himself, on the donkey. Despite the heat we could all laugh at one another, especially over our various attempts to mount.

Hours passed by, and the sand-storm seemed to get worse, so we decided that we should have to stay at Berber for the night, hoping for better conditions on the following morning.

While we were waiting the local chief paid a call, and, as the Commissioner explained to us afterward, the arrival of a flying-boat with, what was most wonderful of all, a woman on board was such an event for him that he could hardly express himself on the subject. He wanted to bestow presents upon us, and insisted that my wife should accept an ox or a sheep. He was prepared literally to kill the ' fatted calf ' that we might feast on this occasion.

We could not but admire the competent way in which the District Commissioner handled the whole situation. We, a party of seven, with a huge flying-boat, were suddenly thrust upon him unawares in an isolated district in the heart of the Sudan, hundreds of miles from any big town, and yet, a few minutes after our unexpected arrival, not only was every comfort provided for us, but every precaution taken for the safety of what was to the inhabitants of the district an extraordinary craft.

Fearing that the current might be too strong for our own emergency anchor I asked if there were any others available, that we might make our machine more secure, and within a few moments anchors were produced. In addition, a guard of soldiers was placed on the shore opposite the flying-boat to take care that no one approached, with orders, should anything unusual occur, to report to us immediately.

Under those happy conditions we went to the Commissioner's bungalow, where, despite the terrible heat, cool drinks were waiting for us, and although the usual dining arrangements were for one, lunch was somehow quickly provided for the extra seven.

My wife and I could not help wondering at the time how the average household in England, in similar circumstances, would deal with such an emergency.

More wonderful still, beds were produced for all our party, and after a happy evening and a good night's rest we were up at dawn, ready to get in the air and on our journey before the heat of the day.

We were all on board and ready to go almost before sunrise, and yet at that early hour the Commissioner came down, shaven and immaculately dressed, to wish us good-bye.

In all my experiences, past and recent, when travelling through the Sudan, I cannot speak too highly of the personnel of the Civil Service of that great country.

After a couple of hours' flight we came upon the place where the Blue and White Niles meet, with Omdurman on the right, and Khartum beyond on the left. With the exception of the Atbara the Blue Nile was the first river to join the Nile since we had left the Mediterranean, nearly 1600 miles away.

We landed on the Blue Nile in front of the Governor's palace, and picked up our moorings. Khartum figured prominently in the African Air Route scheme, as it would have to be one of our main depots. Gladstone and I had a whole host of interviews before us, the most important of all being one with his Excellency the Governor and other members of the Sudan Government.

Our stay was thus crowded with appointments, and that day we lunched at the palace. After staying a full day in Khartum, during which time we surveyed the site for the future seaplane base at the junction of the rivers, we pushed off early in the morning for Malakal, 460 miles farther up the White Nile.

We had quite a good trip, and I was anxious to see how our water and oil temperatures were going to behave. By flying at about three or four thousand feet we found quite cool temperatures, and arrived at our destination just before lunch-time.

We came down fairly quickly from the cool atmosphere of four thousand feet, and it was like going into an oven when we landed on the Nile in the heat of the midday sun at Malakal. As usual, after Worrall had picked up the moorings in the nose of the machine, I switched off the engines and went forward to help him finally tie up. The heat was terrific, and the sudden change of temperature all in a few minutes, from flying at an altitude to the glaring heat of the sun in the open on the river with our craft at rest, was very distressing.

On this occasion we were only a minute on the job,

and I immediately went down into the cabin, which, to my surprise, I found quite cool compared with the temperature outside. Evidently we had trapped and brought down the atmosphere from the cool altitudes above, which had not had time to escape from the cabin or get warmed up. However, these delightful conditions did not last very long, for it soon began to warm up, and we watched the thermometer rise. There was no draught, and as the sun was beating on the metal roof, the interior of our hull, which was so delightfully cool while flying in the air, became like an oven, and during the afternoon registered 135 degrees.

We were now at the north end of the great sudd swamp area, and our next trip was a journey of about 360 miles southward over that more or less impenetrable country of cotton-soil swamp through which the Nile somehow finds its way. There are many theories as to the cause of this swamp, known to the world as the sudd. Strange as it may seem, the current of the Nile as it flows through the swamp is very fast. Perhaps this, added to the fact that the river flows through a cotton-soil area, accounts for the fact that flooding is the immediate result when the waters rise a little.

The sudd swamp area is about four hundred miles square, but for many months in the season fully two-thirds of this area consists of hard-baked earth with huge fissures and cracks in it, caused by the sudden drying of the cotton soil.

We looked upon our trip from Malakal to Mongalla as one of the big test flights of the trip, since we expected to touch our record for heat either at Malakal or Mongalla. Possibly Mongalla has a higher average all-the-year-round heat than any other place on the African air route, as it remains constantly between 90 and 105 degrees throughout the entire year. Farther north, in Southern Egypt or Northern Sudan, at places like Dongola or Merowi, winter temperatures of 70 degrees would be enjoyed, whereas in the summer-time we should have to be prepared to take off and land when the shade temperature was anything from 110 to 120 degrees for many months, while as for the sun temperature, it is best not recorded. However, there is one great thing to bear in mind as far as we aviators are concerned, and that is, that while persons following the ordinary means of locomotion are dragging themselves over the earth's surface and stewing in unbearable temperatures, the air route is passing along five or six thousand feet above the Nile in a delightful temperature of about 70 degrees.

Our journey from Malakal to Mongalla was essentially a flying-boat trip, firstly because it would have been possible to alight at any moment on the waters of the Nile, whereas with an aeroplane, if we had wanted to come down, we should have had to make Mongalla our objective, or return to our starting-point, Malakal, for nowhere over the route would it have been possible to land an aeroplane in safety, owing to the nature of the country. In many places it might have been possible to get down in the dry season, but in the wet period the whole country becomes a vast swamp, and landing in safety would have been out of the question. In these days of civil air transport there should be no reason for a forced landing, other than that caused by the elements, for with the great reliability of aircraft engines, and the fact that they are multiple in a craft, forced landing from a mechanical failure is looked upon as unforgivable. The deciding factor in not using aeroplanes for the Southern Sudan section of the air route would be the fact that in the wet season the existing landing-grounds would become impossible, for their cotton-soil surface would become a quagmire, and the cost of making an all-weather aerodrome from a commercial point of view would be prohibitive for many years to come.

We got away from Malakal in good style, the broad reaches of the Nile affording an excellent take-off. Furthermore, the bend in the river at this point makes it possible to take off in almost any direction of the wind. A few miles above Malakal the Nile turns at right angles, westward, and at this point it is joined by the river Sobat, which comes in from the east, and with its smaller tributaries covers that portion of the Sudan which is inhabited by the tribes known as the Nuers. These people are perhaps one of the most primitive races in the world, wearing no clothing whatsoever. This applies to both sexes, and this is somewhat strange, as in native tribes who do not clothe themselves the practice of complete nudity applies only to one or other of the sexes. The Nuers are generally looked upon as a troublesome tribe, who take a delight in robbing the neighbouring tribes of their cattle, and are generally treacherous. However, by peaceful penetration and careful administration, these backward people are becoming more enlightened, although the process must, of necessity, be slow.

Soon after taking a westerly direction up the bend of the Nile we came to the Bahr el Zeraf, a kind of natural canal which joins the waters of the main body of the Nile some hundred miles farther up, thus cutting off a huge corner made by the Bahr el Jebel, which is the main waterway of the Nile through the swamp. Both waterways are navigable, although the Jebel, being the larger, is used by the big steamers. We followed the Zeraf, wishing to make the shortest route, not that it was necessary to follow any particular landmark, for we were quite confident of hitting our objective, Mongalla, on a compass-course, but we wished to study the Zeraf from an air-route point of view.

After about an hour's flying, away to the west we could spot a silver gleam on a hazy horizon, and soon we found that we were converging once more with the main waters of the Nile. The whole country on all sides appeared to be dead flat, tractless, and featureless, save for the gleaming waters of the Bahr el Jebel and the Bahr el Zeraf, which converged at a point ahead of us, although on approaching we found that the actual joining up of the Zeraf and the main Nile was artificial, there being two canals cut.

This work had been carried out by the Irrigation Department in an endeavour to get the overflow of the Nile away down the Zeraf.

IN A FLYING-BOAT

All around us we could see vast swamp fires caused by the dry papyrus and long grass burning. This happens every year in the dry season, and at times, although we were flying at 5000 feet, we were in a complete fog when we passed through the drift of any particular fire.

We were now flying over a region of impassable swamp, which incidentally is famous for the huge herds of elephants which roam about, unmolested by man, in this impenetrable domain. In the region below Bor there roams a herd that is considered one of the largest in existence, but on my previous flights over this territory I had never been fortunate enough to see them. As we approached this region once more I was hoping that we should be more lucky and that my cinematographer would be able to get what I had coveted so long—namely, an aerial cinematograph picture of this vast herd of elephants.

The great difficulty of spotting elephants in the swamp from a height of 5000 feet is due to the fact that they are exactly the same colour as the swamp itself. However, Gladstone, who had a pair of binoculars, thought that he spotted some in the long grass, and a few moments later I realized that we were looking down on the famous herd. I left Worrall in charge of the controls and went down to give Bonnett instructions about taking the pictures. My crew as a rule were most phlegmatic regarding our aerial cinematography work, but, for some unknown reason, on this occasion they were all excited and most interested in the work of photographing the elephants. When I had given full instructions to Bonnett I climbed into the cockpit once more, and with a nod to Worrall took over control, starting to bring the Singapore flying-boat down in a glide toward the elephants. As we came down with our engines throttled right back we soon passed out of the cool air of 5000 feet to the burning atmosphere of the lower regions. Finally we were in a straight glide with the sun behind us, making for the centre of the herd. Then, fearing that it was unwise to glide too far with my engine cut off lest some of the plugs might oil up in the great heat, I just gave both engines a burst of throttle. We were several hundred feet up, but the roar was fatal, for instantly the whole herd stampeded in all directions. They must have been a thousand strong, and in a few seconds we were right down on top of them, but those few seconds were too long, for in that time they had split up in all directions, so that Bonnett was at a loss to know which particular little group to take. In a few moments we had zoomed into the air again, and I was circling and trying to consider what our next shot would be. In the meantime my crew, for the first and only time on our cruise, seemed to forget themselves. I think that the figure of Bonnett standing by the camera doing nothing, and at a loss to know what he should endeavour to take, was too much for the rest of the crew. The sight of hundreds of elephants stampeding in all directions seemed to stimulate such interest in the work of cinematography that they all started to shout at Bonnett, giving individual orders as to which

106

particular bunch of elephants he was to take. One was shouting, "Over here on the left"; another, "Bonnett, quick, over here on the right. Those great big tuskers"; whereas I was endeavouring to divert his attention to a wonderful little bunch right ahead' of us, toward which I was steering the machine. By this time Bonnett was turning wildly on anything, irrespective of whether he was taking elephants or not. Our real trouble as far as taking cinematographic pictures of such a subject was due to the fact that, owing to the size of our craft, which took nearly a mile to turn, we could not circle quickly enough in order to keep our objective in the picture.

I shall always have distinct memories of those various little groups of elephants, upon which we continually swooped in an endeavour to get some close-up pictures. The main herd may have gone into four figures. Then the elephants scattered in all directions, in little groups of about ten to twenty, and on each occasion as we came down over their heads we could see them frantically struggling to escape from this monster in the sky that for the first time in all history had invaded their domain. They could make but little progress, as they were wallowing belly-deep in swamp, and when we overtook them for the most part they stood still, flapped their ears violently, put their trunks into the air, and snorted with rage.

In every instance, as we swooped down, we met with magnificent defiance from these wonderful beasts, the majority of whose white, gleaming tusks were so large and long that they seemed to go right down into the swamp itself. Great hunters to whom I have talked of these elephants have, on hearing my description of their tusks, usually gone into ecstasies of enthusiasm, for the herd we disturbed must undoubtedly be one of the finest in the world.

For the benefit of these same gentlemen I also plotted on the map the exact position where I located the herd, and they were a little chagrined to think that they would never be able to get near the elephants owing to the fact that they were in the middle of a perpetual swamp country, scores of miles in extent, unapproachable by land or water. However, there were others who rejoiced in the fact that here was a spot where nature would be left untouched, and that these glorious beasts would continue to roam as they had done through the past ages, unmolested and unconquered by man.

The fiasco of our trying to get pictures of these elephants was brought to a close by an incident that might have proved disastrous. Captain Worrall, usually the most composed of men, for the first and only time on our flight was roused to a state of excitement. These hundreds of elephants proved too much for him, and the idea that Bonnett should fail to get pictures on this great occasion seemed to enrage him. I was frantically trying to attract Bonnett's attention toward one particular group that we were circling round, and Worrall, seeing my inability to do so, stood up in his pilot's seat and started to wave his arms in a wild manner to Bonnett, who was in the tail of the machine.

IN A FLYING-BOAT

I was taking the machine round in a circle, trying to keep the sun behind us, so that Bonnett would not get it into the lens of the camera. Just at this moment Bonnett looked round, and Worrall was successful in attracting his attention, whereupon Worrall started to point furiously toward the particular bunch of elephants. Bonnett did not seem to understand, so Worrall, who was looking backward, thrust his arm full out in an endeavour to point out the herd, forgetting for the moment that our giant propeller was revolving at 1000 revolutions per minute, only a foot away behind our cockpits. The result was that Worrall put his finger in the propeller. The impact was so violent that he dropped down on the seat on which he had been kneeling. Fortunately the blow had been a glancing one, and had not broken his wrist or finger, but he was bleeding badly as he crept down into the cabin below. Here he was received by my wife, who, in the sweltering heat caused by our low flying, endeavoured to render first aid. Poor old Worrall was severely shaken, and had to lie down on one of the bunks. I believe that my wife and the rest of the crew were extremely glad when we left off trying to take pictures and climbed away and headed on our journey once more.

Green, our engineer, while endeavouring to assist in the dressing of Worrall's hand, took up a new line of thought, telling the sufferer that he ought to think himself very lucky that his hand had not sent the propeller flying into a hundred pieces. He finished by remarking that anyhow Worrall would know where the propeller was next time.

There actually was a great danger of the propeller flying to pieces. Running into a flock of birds has proved fatal on more than one occasion, for a sparrow has been known to break a propeller by flying into it when it has been revolving at such a tremendous speed. There is no rule in this particular situation: it depends solely upon the circumstances and the angle at which the propeller is hit.

In under an hour Mongalla was in sight, and soon we were circling over this Government station, which is more or less in the centre of the Mongalla Province. Mongalla practically marks the finish of the desert and swamp area, for we had followed the Nile for over 2500 miles through an unrelieved, flat country, and after all these miles we were only 1500 feet above sea-level. It is a very slight fall for the Nile in so great a distance, but now as we approached Mongalla we saw, for the first time since we had left the Mediterranean, a mountain range ahead of us, and a general breaking up of the great plain.

Captain Gladstone, who had been up the Nile before in a seaplane, considered Mongalla the most difficult place of all at which to land a flying-boat on the Nile, so it was with great caution that I glided down toward those dead smooth, oily, and glassy waters. Despite the surface, which was as smooth as a millpond, the current was running at least at six knots. Having made a good landing into a gentle northerly breeze, down current, I decided, in view of the strength of the river, to turn down wind and taxi upstream in order to pick up our moorings. We approached the white buoy without any trouble, and very soon made fast. The instant this was done I naturally shut off the engines. It was when we came to a standstill that we realized the speed and force of the current, and I remarked to Gladstone that we should have to get another line on to the buoy for safety very quickly, as our temporary mooring-rope did not look as though it would hold us. I happened to be looking at the buoy when I felt a slight jerk, and instantly I realized what had happened—the buoy had broken away from the moorings at the bottom of the river.

In a few seconds we were drifting downstream out of control, at an alarming rate, and our tail-plane and port wing-tip were heading direct for some old barges that were moored along the river-bank. If we hit them at the pace at which we were going I knew it would mean terrible destruction, and with a shout to Gladstone of " Anchor, quick! " I was down in the front cockpit struggling for dear life with our 70-lb. anchor. It was an awkward manœuvre to get the flux of our anchor through the narrow gun-ring, and then to balance it on the edge; to erect the collapsible stock, and to put in the key; but somehow, in that sweltering heat of 105 degrees in the shade and possibly 150 degrees in the sun, I struggled with this terrific weight, and performed the operation in a matter of seconds. Our handy man with ropes on such occasions as this was Worrall, but at this time he was lying in agony on one of the bunks with his smashed finger, and I doubt whether I should ever have got the anchor out in time had not my wife sprung into the breach. On her hands and knees in the bottom of the hull she helped me to worm out the anchor through a difficult passage up through the gun-ring. Gladstone, standing on deck, was able to give me a hand in throwing it overboard, and with our eye on the tail-plane and the ever-nearing barges we paid out the line in order that the anchor should get a good hold. Then when we felt we could let the Singapore go no nearer we put a couple of twists round the bollard and waited. There was a breathless moment to see whether the anchor would hold. Then gradually we slowed down, and our wonderful anchor was seen to be holding us fast against the six-knot current.

We were rather proud of this anchor, which was the product of Felixstowe Air Force Flying-boat Base, and the result of much experiment on the part of Flight-Lieutenant Cross, who had presented it to us before we started. Although it weighed only 70 lb. I believe it had the holding capacity of an anchor five times its weight; in fact, it was a proof that weight is not the only necessity for a successful anchor.

As soon as we got ashore at Mongalla we were received by that amazing Commissioner, Major Black, who always seemed to be abounding with energy and efficiency, despite the fact that it was 105 degrees in the shade, day and night, and that there was not a breath of air moving. Already we found our fuel awaiting us on the river-bank, and we went right ahead with that particular job. In the meantime Major Black had found us another mooring, while Worrall had been taken to hospital to have his hand dressed. At the hospital Worrall discovered that his trouble was nothing at all compared with the troubles of two natives who were also there—a mass of wounds, and fighting death. The native of Mongalla is not industrious, and more or less despises work, although it is admitted to be necessary that the women should do a little, such as drawing water from the Nile, and possibly growing the bare necessities of life. He likes to think that his job is to fight and defend, and although the days of war may be over, their pluck and grit still remain, for on the night before our arrival three braves had set out, with spears only, in an endeavour to kill a lion which for weeks had been stealing from their flocks. They cornered and attacked the lion, and a most terribly bloody struggle ensued. They were successful in killing the lion, but at an awful cost, for the beast had killed one of them outright, and the other two were helpless masses of wounds where they had been mauled. Although they made a brave struggle for life they died in hospital the following evening from loss of blood.

We were up before dawn on the following day in order to try to get off the water and attain our altitude before the heat of the sun came up. We got away with ease, and it was a great comfort when, after about fifteen minutes' flying, and with Mongalla a long way behind us, we had got into cool air once more at an altitude of a few thousand feet. We were soon passing over Majuba, the site of the new Government station for the district, and took aerial photographs of the bend in the river, for it is anticipated that this will be the future flying-boat base for this part of the Sudan. It is understood that Redjaf, which is a few miles farther up, will be moved *en masse* down to Majuba, which is supposed to be a far better position for a station. This should be a great junction in the near future, as a big arterial road will link up with the Northern Congo, and other roads are already running from this point to Kenya and Uganda.

Soon after leaving Redjaf the Nile gave way to a series of rapids, and the country on all sides began to rise round us into various mountain ranges, and so we accordingly climbed higher too. We more or less followed the rocky gorge of the Nile, until at last we came out on top at Nimule.

Here the rapids ended, and away to the south-west stretched the broad waters of the Nile flowing through a swampy district at an altitude of about 2000 feet above sea-level. We were now in Uganda, and we flew up the Nile south-westward to Lake Albert. Instead of going up the waters of the Nile eastward toward the Murchison Falls, we carried on due south along the lake shore, until we came to the sheltered little bay at Butiaba, where we landed.

Chapter VI

CROSSING THE LINE

BUTIABA was perhaps the most inaccessible spot that we called at between the Mediterranean and the equator. Our fuel had to be shipped by rail from Mombasa to Namasagali, then transferred to the steamer which goes by way of the Victoria Nile and through the swamps of Lake Kioga to Masindi Port, whereupon it had to be unshipped and put on a motor-lorry, and transferred over a rough road to Butiaba, by which time a very high percentage of the cans were leaking or empty.

I was also informed that the sandy spit on which Butiaba is situated on the shores of Lake Albert is full of soda, so that our petrol, which had been buried on the beach in an endeavour to avoid evaporation from the heat of the sun, suffered severely from the chemical action on the tins instead, and of the fuel that had been originally dumped only about 20 per cent. remained. However, we had a big reserve in our tanks on board, and we were not held up for want of fuel.

We were kindly received at Butiaba, and after a hasty lunch took off for Entebbe, on the Victoria Nyanza.

Lake Albert is a little over 2000 feet above sea-level, and as this was our first high-altitude landing with the Singapore flying-boat we were a little anxious and curious to know how she was going to get off from Butiaba, especially in the heat of the noonday sun. Fortunately the temperature of Butiaba was a full ten degrees lower than Mongalla, but if we had not had this advantage it would have made little difference to our get-off, for to our joy the Singapore, with perhaps a slightly longer run than usual, leaped into the air as though she had been taking off from sea-level in the denser atmosphere of the British Isles.

Our route to Entebbe was going to be direct over the land from Butiaba to Masindi Port, which was on the Victoria Nile, and then we were going to follow a course through the swamps until we picked up the Victoria Nile once more near Namasagili. We were to follow this up a series of rapids to its very source at the Ripon Falls, where the Victoria Nyanza has its one and only outlet, and spills its waters over a ledge, thus starting the great White Nile.

Immediately after leaving Butiaba, on our forty-odd-mile trip over the land to Masindi, we had to climb over a steep escarpment of about 1500 feet covered with dense tropical forest. However, there was no difficulty about this, and we just flew straight at it and climbed over it. When we were half-way across our land-jump between Lake Albert and the Nile at Masindi Port we espied a little lake which looked quite a refuge, and reacted in the same way that a small island might do had we been flying over a long sea-passage in an aeroplane. Even so I have always maintained that I would far sooner be in a flying-boat flying over the land than I would be in an aeroplane flying over the sea.

It was while we were on this journey that we had a serious scare. Conway became convinced that we were on fire. He could smell burning, and pointed out the fact to the rest of the crew, who all agreed with him that something was burning. Instantly a search was made all over the machine, but without avail. Worrall and I were unaware of the trouble that was going on inside, and naturally were a little alarmed when Conway climbed up beside us in the cockpit and shouted in my ear that he thought we were on fire. I leaned down and inquired "Why?" and he replied that they could smell burning everywhere and could not trace it. It was then I could see the humorous side of the situation, because for the last ten minutes we had been flying into the smoke-drift of a large forest fire, whose pungent fumes might easily have made one imagine that burning was taking place next to them instead of two thousand feet below.

Soon we were flying over the vast swampy region of Central Uganda, and later passed up the last reaches of the Victoria Nile. The last fifty or sixty miles form a series of little waterfalls, a very beautiful sight from above, and away to the south we could see what appeared to be an ocean, but what was really the Victoria Nyanza, which is two hundred miles square, and whose surface is 4000 feet above sea-level.

Before long we had reached Jinja, and next door were the Ripon Falls, the source of the great White Nile, discovered by Speke in 1862. Speke's journey from the east coast, up round the shores of the Victoria Nyanza to Jinja, although only some 600 miles, took him nearly three years on foot, his delay being caused by the various chiefs who detained him, more or less against his will, from time to time along the route.

From here we turned westward, and in the afternoon sun we flew along the northern shores of the lake, until Entebbe was reached. Before leaving England I had been cautioned by many folk to take care in landing on the Victoria Nyanza. Not only is it over 4000 feet above sea-level, but the relative density of the atmosphere is such that under certain weather conditions on a very hot day the atmosphere at lake-level has a density equivalent to that of the air at 10,000 feet in the British Isles.

Therefore when we came down to land, although I had no actual fears, I took every precaution that there should be no mistake about the job, and in order to allow for the possible thinness of the atmosphere I landed fairly fast, with the result that I had to hold her off while we lost flying speed, our hull making a clattering noise as it skimmed over the little wave-crests of the short, choppy sea that was running before the afternoon breeze.

We had no difficulty in picking up our moorings,

and our arrival being a little unexpected the native population was more or less taken unawares. We had a wonderful reception, and were soon drinking tea on the lawn of Government House, where, under the shade of tropical foliage, we looked down through giant trees to the blue waters of the lake, one of the finest vistas of scenic splendour I can remember on any of my long journeys across the world.

It was here that we had to carry out for the Colonial Office the experimental flight with our flying-boat, that they might know of the possibilities and difficulties of an air-mail service by way of the Nile to the Victoria Nyanza. Therefore it was agreed that we should write a report on our experiences during our flight from Alexandria to Mwanza, on the southern shores of the Victoria Nyanza, whereupon we should about turn and fly all the way back to Khartum, and again return to Mwanza, thus completing the experimental test journey.

I decided that as I should have to return to Entebbe it would be a good thing to fly down the east side of the lake *via* Kisumu to Mwanza, and return up the west side by way of Bukoba, thus making a circumnavigation of the lake. I extended an invitation to the Governor, Sir William Gowers, our host, to accompany us on our journey round the lake.

He was very much taken with the idea, and so two days later we took off from Entebbe and flew round to Port Bell, some thirty miles away, where the Governor and the party boarded the Singapore.

The main idea of this manœuvre was to test out

Port Bell as a flying-boat base, which incidentally was up a sheltered inlet and proved to be most satisfactory in every way. Port Bell is the port for Kampala, the commercial centre for Uganda.

We were soon all aboard and had taken off *en route* for Kisumu, Kenya's port on the lake, 150 miles away.

Now this little journey was rather a famous one, because we were going to cross the equator, and we discovered that it was going to be one of the "first occasions in the world's history," since it would be the first time that a woman had ever crossed the equator in a flying-boat. As a matter of fact, my wife had never crossed the equator before, and so it was rather an added attraction that she should be doing it in an up-to-date style and at the same time creating a record.

Sir William Gowers was anxious that we should celebrate the occasion suitably. Aviation being still in its early youth, the air had no traditions and customs like those of the sea, and we could not think of any figure corresponding to Father Neptune. Icarus could not very well descend from the skies, and so we finally came to the conclusion that the best way to celebrate the occasion would be to toast ourselves in champagne at the precise moment that we were crossing the line. There was going to be no mistake about this, for it so happened that there was a little uninhabited island in the lake whose southern-most point, according to the map, was right on the equator. Therefore I headed and flew straight over the southern shore of this island, and at the precise moment that we crossed the line, the corks being drawn and our glasses filled, a toast was proposed by Sir William Gowers, first to Lady Cobham on this great occasion, then to the success of the future African air route. Poor old Worrall was out of this, having been left in charge of the controls in the cockpit. However, after I had partaken of some of the lunch that followed, I went forward and relieved him, while he came down below, had his repast, and finished up with liqueurs and cigars.

After all that we had heard about the difficulties of flying round the Victoria Nyanza I came to the conclusion that it was one of the most delightful places in the whole world for flying, and this impression was strengthened by succeeding flights. Always there is extraordinary visibility, the rainstorms are invariably local and can be dodged, and as for bumps, we never experienced any. In fact, our luncheon-table was far steadier than any train, and quite as firm as any ship. Our cooking apparatus consisted of a Primus stove on top of a bench, and it is interesting to note that the stove was in no way secured to the bench or the utensils to the stove, and never once on our 23,000-mile journey round Africa did we have the suggestion of a spill.

We arrived at Kisumu, at the far end of the Kavirondo Gulf, some time after midday, so that when we landed we glided down into a stiff breeze off the lake, and finally alighted on a very rough, although short, choppy water. Most days the lake is dead calm up to about noon, and then a breeze seems to rush in off the lake and continue to blow in this manner until about sunset. The rough sea in no way upset our Singapore flying-boat because the wave-crests were so short. Nevertheless it prevented us from refuelling that afternoon, as we could not get a small boat alongside to hand the heavy drums of fuel on board without running the risk of possibly damaging our flying-boat by bumping her severely in the rough sea.

That night the Governor of Uganda stayed with the District Commissioner of Kisumu, and we were fortunate in being of the house-party, the rest of the crew being distributed round the town and looked after by many kind folk. It was at Kisumu that we said good-bye to Gladstone, who, the moment we arrived, made a sudden dash to catch the train that was leaving for Nairobi. He had but a few minutes in which to do this, and in the rush left various oddments of his extensive baggage behind.

On the following day we continued our journey down the eastern shores of the Victoria Nyanza as far as Mwanza. A more delightful trip I cannot imagine, and toward the end of the afternoon we were circling over that beautiful harbour, with its picturesque rock-formations jutting up on all sides, and the silver-sand shores bordered with the most wonderful palm-groves.

It was at Mwanza that we were nearly sunk while at our moorings on the water. A party had come out to greet us in a very ancient steam-launch, which some said would soon be qualifying for its centenary. The launch was crowded with people, but the skipper wished to come on board our flying-boat, so after stopping his launch a little way off he left a native boy in charge and came alongside us in a skiff. In the meantime, from another angle, we were being approached by a kind of steel tank, which proved to be an old pontoon-section. The natives who were paddling it had little or no control, and I could see that if they touched us they would drive a hole through our thin metal hull.

I was shouting to them to go steady, when suddenly there was a cry from Lady Cobham, " The launch is drifting on us!" I turned to find that the old steamboat was indeed out of control and drifting right on to us.

I shouted to those on board to go astern, and the poor wretched native in charge, feeling that he must do something, unwittingly put her " full steam ahead." It was a terrible moment, and it seemed as though disaster could not be averted, for the skipper of the launch was on board our craft at the time, and was frantically trying to shout instructions.

However, just as the crash seemed inevitable, one of the passengers on board the launch acted upon the shouted exhortations of the skipper and pushed the lever into " reverse." It seemed as though it would never pull up in time, and the heavy iron bows were within an inch of our hull when she started to go backward, and we were saved.

Stiebel, the Commissioner at Mwanza, proved to be

an old friend, for although we had never met before we had corresponded much together in connexion with my arrangements for the preparation of the landing-ground at Tabora, on the occasion of my first flight from London to the Cape. Stiebel was Commissioner at Tabora at the time, and went to no end of trouble in making the Tabora site suitable for me to land on. However, we never met on that occasion, because when I eventually arrived in Tanganyika on my first flight Stiebel had gone home on leave, and that entertaining gentleman Mr Buckley was deputizing for him.

Stiebel, aided by Mrs Stiebel, proved to be a wonderful host, and that evening, as we sat on the veranda of their house on the side of the hill, which faced north-west and the sunset, we had a marvellous vista of the bay below us, surrounded on all sides by weird, fantastic rocks piled on one another, balancing at perilous angles.

Wonderful avenues bordered by mango-trees led from the shore to the various houses of this little colony, who had given us such a wonderful reception and open hospitality, our crew being taken care of by various members of the community.

The next morning we had to say *au revoir*, for now, according to the arrangement with the Colonial Office, we had to turn about and fly all the way back to Khartum. This was a somewhat gruelling procedure to those of us who were actually carrying out the flight, when we thought of the miles that we had to cover right down to South Africa and the long homeward trek up the west coast. However, the journey had to be done, and so the next morning we were early aboard. After casting adrift we had to taxi quite a long way out of the bay into the lake, in order to get a clear run head into wind. This was just one of the little snags of the sheltered cove.

We were soon off and in the air with a run of some thirty seconds over the water, and flew along the southern shores of the lake amid delightful scenery. Then, meeting one or two tropical showers, we easily avoided them by altering our course a little, and crossing their path before they reached us, or going behind them after they had passed.

Later we flew up the western shores past Bukoba, and then took a direct course for Port Bell, passing right over the Sesse Islands, which in the past were so severely attacked by sleeping-sickness that the authorities considered it best for the natives to evacuate, and so for the time being the islands were uninhabited.

By noon we had arrived at Port Bell, and here the Governor went ashore immediately in order to keep an appointment in Entebbe, some hours' motor-drive away. We had no moorings laid down specially for us at Port Bell, as we meant to make that arrangement later in the afternoon, and in the meantime contented ourselves with dropping our own anchor in those calm waters.

Green, our engineer, wishing to make some slight repair to the tropical radiators, had drained the starboard engine of its water, and had quietly set to work upon his task by taking the radiator off. In the meantime Worrall and I had gone ashore to search for some suitable form of mooring, and were being helped in this matter by the captain of the lake steamer, who happened to be in port at the time. Suddenly, from the decks of the steamer, I noticed that the Singapore was gradually drifting ashore, and that the breeze off the lake was gently blowing her toward a dense swamp of rushes and grass. Undoubtedly the Singapore was dragging her anchor, but the seriousness of the situation lay in the fact that away over the hills, only half a mile away, dense black clouds were rolling up, threatening a sudden tropical squall that might descend upon us in a very short time. Now the situation would have been quite different had the Singapore been drifting out into the middle of the lake, for had she then been hit by the gale and storm she would naturally have headed into the wind and ridden it out with ease. But should she be locked hard and fast into the rushes, with her tail into the approaching gale, and unable to turn and head into the wind, the situation would be very serious, because she might run the risk of being turned over.

Worrall and I realized this in a few seconds, and instantly rushed from the ship down to the quayside. There we leaped into a skiff and quickly got on board our flying-boat. We soon discovered that we had dropped anchor into a deep bed of water-weeds, which had held us fast until the wind had begun to blow. Then gradually they had given way, and our anchor was dragging a great wad of underwater growth, so that we were gently slipping back into the rushes. There was only one engine which we could start, and I hoped that with this running I might be able to get out into the lake and anchor into a firm bottom, but for some unknown reason, although our engines had never failed to start up under ordinary conditions, on this occasion we could not get the port engine running. Green, Conway, and Worrall struggled in vain in the heat of the tropical sun, swinging our starting-handles until I feared that they would do themselves some physical injury. It is in moments like this that one realizes the shortcomings of one's craft, and ever after this I shall be an advocate of a satisfactory self-starter. Even when she did fire finally I could only taxi out on one engine at an angle of 45 degrees, whereas with a sound water-rudder I could have gone out under my own power on one of my engines at almost any angle I chose. The seriousness of the situation was averted by the fact that the squall took a sudden turn and did not reach us. The captain and crew of the steamer, too, lent generous aid by dropping a weighty anchor well out into the lake, from which a long line was thrown to us, and we were able to draw up to it, safe against all emergencies.

Sir William Gowers had been our host at Entebbe, and therefore it was with reluctance that I contemplated leaving the comfort of Government House and flying all through the heat of the Sudan to Khartum and back again, a journey of some 2700 miles, and possibly under the worst conditions we were to experience during the whole of our trip round Africa.

However, I realized that to prove the feasibility of flying over this particular section of the route by journeying all the way to Khartum and back again would do so much to influence the East African Governments toward supporting the through-Africa air-route scheme.

There was no reason why either my wife or Bonnett, my cinematographer, should take this journey again, and so it was arranged that Bonnett should stay behind at Kampala and take cinematographic pictures of topical interest, and that my wife, after a few days' stay at Entebbe, should journey up to Nairobi, where I should rejoin her after the completion of our return trip.

It had been suggested that as the Governors of East Africa would be present at Nairobi at the end of the week, in connexion with the Hilton Young Commission, it would be a unique opportunity, after the successful completion of the experimental flight, to discuss their future intentions regarding the African air route.

I was determined to make the flight to Khartum and back in record time on schedule, and so, with an early start from Port Bell, we took off on a non-stop flight to Mongalla. We went more or less direct to Port Masindi, and then, instead of cutting over to Butiaba, we followed the Nile northward, until we came to the famous Murchison Falls, where the Nile tumbles through a deep ravine, and fish and all live matter are hurled pell-mell over the rocks. It is said that the crocodiles congregate on the lower side of the falls in thicker formation than perhaps in any other part of the world. They literally lie about on top of one another. However, from the altitude at which we flew we were unable to distinguish this spectacle, and so passed on to the main stream which flows toward Nimule. Once more we flew above the ravine down which the Nile tumbles some thousand feet until we came to Redjaf. From this point to its mouth the Nile is more or less navigable—a distance of nearly three thousand miles.

After a journey of 400 miles we landed at Mongalla. Again we were thankful for the ready help of Major Black, who had our fuel supplies waiting in readiness, and quickly we refuelled, setting out once more over the great sudd swamp area toward Malakal. On the way just beyond Shambe we passed over an Air Force camp where they were endeavouring to quell a disturbance caused by some troublesome natives. These seemed to take a delight not only in murdering white men who were trying to keep peace in the country by fair dealing, but also in raiding neighbouring tribes, stealing their cattle, and waging warfare in every direction.

Well before sunset we arrived at Malakal, and were soon moored safely, and able to refuel our machine before darkness overtook us. Thus we were ready to set out on the 550-mile hop for Khartum on the following day.

We took off into a strong northerly breeze, which in this part of the world is the prevailing wind. Considering its persistency and strength we realized that this might prove to be a severe handicap to the regular running of the air route on its journey northward.

On this trip we were able to make some interesting experiments and collect some valuable data. Every quarter of an hour on the flight I altered our altitude, flying for fifteen minutes at a given height, marking our progress on the map, and thus calculating our forward speed. I did this every five hundred feet. Thus we flew level for fifteen minutes, going up in steps of five hundred feet until we were at an altitude of 8000 feet.

The cruising speed of our machine in still air was 95 miles per hour, but we had taken off into a 30-mile-an-hour head wind, and thus our forward speed was only 65 miles per hour. We found that the higher we went the more our speed increased, and at 6000 feet we discovered that we were making fully 95 miles per hour, which proved that we were in a region of more or less still air. At 8000 feet, however, we were happy to discover that we were making about 110 miles per hour, which proved that at this altitude the wind had changed to the opposite direction, and instead of blowing from the north it was now coming from the south, at 10 miles or more per hour.

Soon after midday we landed at Khartum and set to work to refuel. That evening our fuel agent in Khartum came to me with a great story about two lady tourists who were desirous of flying over the great sudd swamp area to Mongalla. Would I take them with me on the following day? Now ours was an experimental flight, and we were not out to take the public on joy-rides, but these two ladies, who hailed from the New World, descended upon me in person, and after I had told them of the tedious return journey by boat they still pleaded, so finally I gave way.

It was my intention on the following day to fly without a stop from Khartum to Mongalla, a distance of some 810 miles, and I hoped, with the following northerly breeze, to accomplish this journey in a matter of eight hours. The journey back by river steamer would take something like ten days, but my two lady passengers were determined on the venture.

During the afternoon I journeyed into the native quarter of Khartum, to find a tinsmith who could carry out for us some minor repair. I shall always have vivid memories of that native craftsman in his little old shop, surrounded by pots, kettles, and lamps, a veritable scene from *The Arabian Nights*, for his equipment was at least four hundred years behind the times. A small boy blew with ancient bellows some charcoal into a red-hot glow, thus the soldering-iron was heated. And while the tinsmith was at his job the shoemaker next door lit a small fire in the roadway, and, by means of an old tin contrivance on the spit principle, with two long fingers turned a twisted-wire handle, thus baking his coffee-beans, while the smoke-drift from his fire almost fumigated the tinsmith, who, however, seemed to offer no objection.

The next morning at five o'clock, and just before

the sun rose, we were down by the waterside of the Nile, where we found our lady passengers, who, despite the fact that they had attended a moonlight picnic, were already waiting. As the sun came up over the horizon we left the waters of the Nile. Having climbed to some four or five thousand feet, and headed southward up the Nile once more, I began to attend to my two passengers. During the first half-hour they were all agog with excitement, first at seeing Khartum pass away beneath them, then at the vast views of the desert, with the silver gleam of the Nile running north and south. Presently we gave them a little refreshment, and both seemed to be thrilled at their new experience. The machine was as still as a rock in the air, and being tired of viewing the passing panorama from the various cockpits of the machine the ladies had sat down on one of the bunks and started to read.

Knowing that they must be tired after their early rise, I suggested that they should lie down on two of the bunks, and forthwith my crew soon got two of these improvised beds prepared. They were soon asleep, and seeing they were neatly covered up with rugs we left them to rest awhile, and devoted our attentions to the journey southward.

We made very good progress, and before noon were passing over Malakal. Once more we passed over the sudd, and finally, exactly eight hours after the time of starting, having accomplished a journey of 810 miles, we landed at Mongalla.

The moment we were on the water all our attentions

were naturally directed toward the business of mooring up, and when this had been successfully accomplished we then began to attend to such matters as refuelling and getting out our baggage for the night's halt. It was during the latter process that we discovered we had two ladies fast asleep in our bunks, and for the first time after putting them to sleep early in the morning we remembered our passengers. So peaceful had been the cruise, so seductive the droning of the engines, that they had slept solidly for some six or seven hours, and when we awoke them I hardly knew whether they were annoyed or not at having come some 800 miles by air. Their only aerial memories of the Nile could be the few brief moments of the start at Khartum, and now it was going to take them ten days to get all the way back again.

The next morning we took off and flew to Butiaba again, and after a brief stop journeyed on once more by way of the Nile to the Ripon Falls, thence round the coast of the Victoria Nyanza, and, instead of calling at Port Bell, flew on and landed at Entebbe, the capital of Uganda.

Thus we had completed a journey from the Victoria Nyanza to Khartum and back, a matter of some 2700 miles, in four days, the same journey taking, by the existing means of transport, well over three weeks at the least.

On the occasion of our first visit to Entebbe what natives there were present to witness the landing looked upon the spectacle as the most marvellous thing they had ever seen. They had heard from the

Commissioner about the proposed arrival of our machine, but seeing is believing, and the news of our first arrival spread like lightning throughout the countryside.

Each native as he told the story added a little bit of his own to it, with the result that most weird and wonderful tales were current regarding our flying-boat and its personnel. They had heard that we were made of metal, hence the story went round that the " great white bird " that flew in the air and, what was most fascinating to them, floated on the water was made of iron.

Our quick departure from Entebbe on the first occasion had been a great disappointment to all the natives who were going to journey in from afar to the capital in order to see the wonderful spectacle, but when they were told by the Commissioner that on a certain date we should return they were pacified.

Among these natives were the canoe-men from the Sesse Islands, and when we returned from Khartum we had no sooner got on to the water than we noticed, to our great surprise, that we were being descended upon by an armada of war canoes, with all their crews chanting and singing and bedecked in gala costume.

A grand reception had been arranged in honour of our return, and as we taxied quietly over the water to our moorings we thoroughly enjoyed the spectacle of these giant canoes, each manned by about twenty natives, paddling rhythmically as they pursued us.

Just as we picked up our moorings I noticed that

134

CANOES BEARING DOWN ON THE SINGAPORE

THE WAR CANOES AT REST, ENTEBBE

each canoe had a long spike in front of it, and I recognized these as the famous war canoes of the Buganda. To my alarm I noticed that the natives were getting more excited every moment, and in a few seconds would be descending upon us in wild confusion. In a flash I remembered that their conception of our metal craft was that it was made of iron. Perhaps they thought that it was a quarter of an inch thick, instead of which it was in reality made of duralumin, one thirty-second of an inch thick.

Many of these canoes had been paddling for days, and had journeyed far in order that they might see the "great white bird" that had flown all the way from England.

When I saw this wild medley fast descending upon us in what appeared to be hysterical excitement I could only fear the worst.

We shouted and we screamed, and when it seemed inevitable that they would charge right into us, from only twenty feet away, the mass of canoes divided, passing right and left of us and just skimming under our wing-tips. Then followed an exciting parade, in which scores of canoes raced round and round us, cutting in on one another until I thought that any moment there would be a collision, and they might possibly charge into our wing-tip float and wreck it, or might go too close under our tail and puncture our hull.

The only person who seemed oblivious of the danger of the episode was Bonnett, our cinematographer, who was circling round in and about the medley of canoes in a motor-launch, taking films in all directions.

Eventually the Commissioner blew a trumpet and called a halt, and after about a quarter of an hour our grand reception was over, and we were escorted ashore for the night.

We were enjoying a little dinner-party given specially in our honour, when Green, whose turn it was for watch that night, came up to Government House to tell me that he was very worried about the position of the Singapore. In tying the lamps on the wing-tips, an operation which had to be effected by running out on the lower plane, he had noticed that his weight, when he had run out on the starboard wing, had not weighted that wing down. He then got a native boy to come out on the starboard wing with him, but even the two of them could not weigh this down, and thus lift the port wing-tip float from the water. From this he came to the conclusion that the port wing-tip float must have sprung a leak and was becoming waterlogged.

We all rushed down to the lake-shore as quickly as possible, and were soon clambering on board. We found then that four men by climbing out on the starboard wing were just able to weigh it down and lift the port wing-tip float out of the water. Green, Conway, and I then got into a small skiff and rowed under the port float, and after many struggles, bobbing up and down in a slight chop, Green finally managed to insert his screw-driver and get the drain-cock undone. We then discovered that our surmises had been right, for immediately the water gushed out, and furthermore there was a tell-tale little stream from the keel of the float, which showed where the rivet had come out, thus causing all the trouble. We had to hang on in this position for about two hours until the float was finally drained. The humorous thing was that we were all in our dinner-jackets, which were becoming somewhat bespattered in the adventure. This little episode made me realize the necessity of having floats made into several water-tight compartments or bulkheads, so that, should a float spring a leak, the flooding would be only local, and instead of filling the whole float with water the danger would be averted, as the one little compartment that was flooded would not be of sufficient weight to submerge the float.

One of the ready volunteers to help us on this occasion was the secretary at Government House. Just as we were leaving the craft one of our most useful lines was dropped overboard. We shouted to Conway to grab it, but he missed. The secretary then quietly asked me, " Do you want the line? " and I replied, " I don't know how we can replace it." Whereupon he calmly stepped overboard into the lake, complete in his evening attire, and after disappearing out of sight beneath the water emerged almost immediately, having retrieved our cherished line.

This same gentleman was a renowned elephant shot, and the story goes that on one occasion, when out hunting, he got about fifty yards ahead of his party, when suddenly an elephant charged down upon him. He raised his gun to take aim. Just at that moment one of the party in the rear, who could not see the elephant, shouted a question to him, whereupon the secretary, with the elephant towering above him, looked over his shoulder and said, " Just excuse me a moment," and then, taking aim once more, brought the elephant in a heap at his feet. On being questioned afterward as to where he hit the elephant, in such a desperate moment, he simply replied, " Oh, just the brain shot."

Chapter VII

KENYA AND NYASALAND

THE next day we flew to Kisumu, where I found Mr Carberry waiting for me. He had flown down from Nairobi with Captain Tymms, the Air Ministry representative, having come down specially to fly me back to Nairobi to attend a Governors' Conference that was going to be held there. This had been convened in order to make a decision regarding their next action in supporting the African air-route scheme. It was a favourable opportunity as the Governor of Uganda and Sir George Schuster, representing the Sudan, were already staying with the Governor of Kenya, in connexion with the Hilton Young Commission. Mr Carberry can be looked upon as one of the pioneers of civil flying in Kenya, and I considered it a great compliment, as well as a kindness, that he should have flown about 200 miles down to Kisumu in order to take me back to Nairobi.

Accordingly I left Worrall in charge of our flying-boat at Kisumu, with a whole list of odds and ends to be done, and Green also was anxious to have a thorough look at his engines. We took off from the aerodrome at Kisumu to climb to the heights of Nairobi. We had left Entebbe at about 7 A.M., and had done our 150-miles flight to Kisumu in well under two hours, and before ten o'clock we were in the air, climbing away eastward to mount the first ridge. The aerodrome at Kisumu was 4000 feet above sea-level, Nairobi is 6000 feet, but it is necessary to climb to about nine or ten thousand feet to clear the first escarpment well. I was very interested in the journey because it will be the line of the future air route. We did not land at Nakuru, which is about half-way, but continued on, passing over the dry arid region of the great Rift Valley until at last we landed at Nairobi.

Here I found my wife waiting to meet me at the aerodrome, for not having been down with us on the return journey to Khartum she had travelled up by train from Entebbe to rejoin me at Nairobi. Here I met Gladstone, and after an evening's consultation we made our plans for a conference on the following day.

I firmly believe that the East African Governments were convinced of the utility of air service, inasmuch as the conference was a complete success, and we received definite promises of help.

It was while we were at Nairobi that the first flight in history was accomplished from Mombasa to the capital of Kenya, and furthermore it was done by a woman, Mrs Carberry. She had learned to fly in England, and had brought her Moth aeroplane out by ship to Mombasa. Having unpacked it on the quayside, she flew with a passenger up to Nairobi.

The job ahead of us now was to complete our flight of survey round the African continent, for then I felt that we should be in a better position than

anybody else to deal with the air-route scheme,
because, firstly, we should have surveyed all possible
routes through Africa to Europe, and furthermore we
should have tried them out on all kinds of craft.

In 1925 I had flown from London to Cape Town
and back in an aeroplane, then there had been the
various Air Force flights by aeroplane, all of which
gave us excellent data for the air-route scheme. My
partner, Gladstone, had flown up and down the Nile
on a seaplane, so that we knew all about the route on
that particular craft. And now we were going through
on a flying-boat, and furthermore we were going to
return *via* the west coast. I felt in consequence that
on the completion of our journey we should be in a
sound position to make correct decisions as to the way
in which the job ought to be done.

Already I was of the opinion that it was essentially
a flying-boat route from the Mediterranean to the
Victoria Nyanza, and an aeroplane route from the
equator to South Africa.

It was arranged that Gladstone should go on down
into Tanganyika in order to visit the Government at
Dâr-es-Salaam, and Captain Tymms was going to
look over a line of aerodromes that had already been
surveyed by the Tanganyika Government. So here
we parted, and my wife having gone ahead by train
to Kisumu, Carberry once more kindly flew me down
in a Moth, back to the Singapore, which was now all
polished up and ready for continuing the journey.

Before leaving Kisumu there was one thrilling
incident that I must relate—namely, the reception

given to us by the natives on our return to the aerodrome. They had journeyed for miles for this special occasion, and had bedecked themselves in all their wonderful feathers. Their headgear was the most decorative form of headdress I have ever seen, and would put the creations of our most famous revue artists to shame. The *pièce de résistance* of a London or Paris revue would have given untold wealth for the headgear of one of these natives. It consisted of three banks of white ostrich feathers piled one above the other, each hat containing possibly a hundred perfect plumes. Their bodies were covered with magnificent leopard-skins, and very soon there were scores of men dressed in such attire dancing round our aeroplane in great circles.

Then came a procession of such gentlemen riding on bullocks, followed by another party whose headdresses consisted of rows and rows of rhino tusks, and for nearly an hour they chanted and sang and danced around our machine in order to give us welcome.

Most of them had painted their limbs, the chief vogue being to have one white leg and one blue leg, while others had two red legs and white arms, and often there were weird designs in red, blue, and yellow painted on their faces. It was an extremely hot day, and it was just past noon when this ovation started. The terrible exertions of their dances must have been most exhausting under such conditions; in fact, they all seemed to be in a bath of perspiration after a very short time, but I soon discovered that the hotter they became the more they seemed to love it.

IN A FLYING-BOAT

That afternoon we returned the natives' call by visiting two villages near by that were inhabited by the natives of the Kavirondo Gulf. In one little village we were introduced to what they told us was a great-great-great-great-grandmother. They said she was one hundred and six years of age, and had six hundred descendants, and, as far as I could understand from the native interpretation, the six women and children to whom we were introduced afterward represented the seven generations. We took a picture of them and christened it " The Seven Ages of Woman." At another village we discovered that it was the custom to tattoo one's tribal history on the body, while there were others who had earrings that weighed several pounds. I felt the weight of them myself, and noted that the lobe of the ear was weighed down seven or eight inches in consequence, for these ornaments were permanent fixtures and could not be removed.

That night we made preparations for our departure early on the following morning, when we were going to continue our journey in all earnestness after our success in East Africa.

Once more we arrived in Mwanza, and were again greeted by Mr and Mrs Stiebel. Our journey from Mwanza to Kigoma needed careful thought, for after flying along the southern shores of the Victoria Nyanza we were then going to fly over a land of uncharted mountain ranges with a flying-boat, until we came to Lake Tanganyika, where our destination was to be Kigoma, some way down its eastern shore.

The map on inspection appeared to be full of

detail, and of apparent accuracy, but we were soon to find out that this was not so.

On our second take-off from Mwanza our flying-boat was fully loaded, and I shall always have vivid memories of that brilliant sunshiny morning when, with our backs to the sun, we opened out the throttles, and were soon racing over the rippling blue waters on our last take-off from the Victoria Nyanza. Gradually we left the water and headed westward.

It had been my plan, after leaving the lake, to fly on a course that would take me in a series of jumps of fifty miles each, more or less, along a chain of small lakes that were marked on the map, with a final hundred-mile jump over a mountain range marked at 6000 feet that would bring me out at the north end of the Tanganyika Lake.

Not long after leaving the Victoria Nyanza behind us and passing inland I discovered that the country did not correspond with the map. Where small lakes were marked we found, on arrival at the scene, no sign of any water whatsoever. However, allowing for some inaccuracy in our chart, I continued on a compass-course and finally recognized one of these small inland waters.

Ahead of us we could see rolling uplands with a mountain horizon towering above us, and considering that we were already flying at 6000 feet I felt a little alarmed about the correctness of our course. However, after our previous experience of inaccuracy, I came to the conclusion that the marking of the contours on our map was also untrustworthy, so we decided to

SIR ALAN AND WORRALL WITH SOME NATIVE FRIENDS AT
KISUMU

SEVEN AGES OF WOMAN   144

depend solely upon our compass. We knew that if we could climb above the range ahead of us we could not easily miss Lake Tanganyika, considering its size, so we comforted ourselves and flew on.

I do not know who made the map; perhaps it was done from memory. As we continued to climb, and still the mountains got higher and higher, I came to the conclusion that quite possibly the fellow who had made the map had based it upon the report of a friend who had known some one who had been there. Finally we found ourselves at 10,000 feet, but even so ahead of us were ugly jagged peaks with clouds hanging on top of them.

About this time I noted the top end of a deep valley which ran away to the south-west, and after calculating the time that we had been in the air, in relation to our air speed, I decided that we must now be flying over the approximate watershed, and as there was a stream in the bottom of the valley I was more or less assured that it ran down into the Tanganyika Lake.

We started to follow the course of the valley, when we came to a cloud-layer, and fearing the difficulties of climbing above this cloud, and the possibility of the clouds mounting higher and higher until such a time that our machine should have reached its ceiling, and would be unable to surmount them, I decided to go beneath the cloud down the valley.

Soon we were flying down a kind of tunnel, with the rock-walls of the valley on either side and the clouds as a roof.

The only thing which assured me that we were not

going to fly into a *cul-de-sac* was the fact that the river was getting bigger and bigger, and I knew that it must come out into the main lake.

We were all very happy when we saw the blue waters of the Tanganyika ahead of us down the gorge, and soon we were flying out over the centre of that great lake, which is 400 miles long.

Having made a very early start we were able to come down and land in the little bay at Kigoma by noon, where excellent moorings and facilities were awaiting us. That afternoon the Deputy Commissioner very kindly drove us over to Ujiji, near by, and we saw the famous mango-tree under which Livingstone was sitting, exhausted and ill, when Stanley came upon him.

It was here that I came to the conclusion that, after all, the ladies were the mercenary sex. We wished to take a picture of a little native girl with her neat headdress and costume, but even in the heart of Africa she could not consent to stand for us unless we paid her a shilling. We wanted the picture, so gave in, and afterward christened her " Mercenary Mary." The great point was that until she had actually received the shilling we were not allowed to take the photograph.

That night the Deputy Commissioner had arranged a special party and dance in our honour at his wonderful residence that stood high on the hill over-looking the lake. There was a magnificent broad veranda that was lit by a full moon, and the whole made a very picturesque setting.

IN A FLYING-BOAT

I believe the building had been originally built as an hotel, but, that scheme having failed, the house has been used as the Commissioner's residence since the British took over the territory.

When flying from north to south of Africa one has to pass through the rainy belt somewhere, and the place varies according to the period of the year in which one is flying. Thus in February and March we met the rainy belt about 500 miles south of the equator, so that our journey from Kigoma to Mpulungu, at the extreme southern end of the lake, was a most interesting flight.

We encountered a series of tropical storms, but discovered that they were all very local, and by making a more or less zigzag course from one side of the lake to the other, which was often only thirty miles wide, we managed to dodge between them. To fly through any of these storms would have been extremely dangerous, I think, apart from the difficulty of such a manoeuvre, because their intensity was so great that they looked like a blue-black mass floating across the sky, pouring down a gigantic waterspout, and all the time forked lightning flashing earthward.

They were veritable ' monsters of the elements,' and I was only too thankful to find that they were local and could be dodged, otherwise they might prove a severe hindrance to the regular running of the future air route.

I suppose that Mpulungu, our next stop, was one of our loneliest ports of call throughout the entire flight round Africa. Mpulungu is the name given to a place situated at the extreme southern end of the Tanganyika Lake. It is a few miles over the border of the Tanganyika territory, and comes under the administration of Abercorn, which is the most northerly province of Northern Rhodesia.

We landed in calm waters between an island and the mainland, and after we had got moored up we noticed a motor-boat pushing out from the shore. I soon recognized the Smith brothers, whom I had met at Abercorn on my previous flight. They are old pioneers in this part of Rhodesia, and had travelled many miles from their estate to await our arrival.

It is interesting to note that the Smiths—with the exception of Mr Venning, the Commissioner, and his assistants up at Abercorn, on the plateau, some fifteen miles away—were the only white men for a hundred miles in any direction, indeed, for two or three hundred miles on many points of the compass. Furthermore, the native population is very sparse in this part of the world.

The spot where we landed was more or less free from tsetse fly, although the wooded island that rose like a sugar-loaf out of the water in front of us, we were told, was full of sleeping-sickness, and was uninhabited in consequence. I feared for the safety of our crew who were working on the Singapore, which was anchored half-way between the island and the mainland, but I was told that the tsetse fly never goes in for long flights on his own, and furthermore never comes out in the sunshine, but keeps in the shadow of the bushes. If he does do any travelling he rides

AN ASSISTANT DISTRICT COMMISSIONER'S OUTPOST,
TANGANYIKA LAKE

ANCHORED AT MPULUNGU, WITH THE SLEEPING-SICKNESS
ISLAND BEYOND

as a rule on the backs of natives or animals, so that, although the flying-boat was only 200 yards away from the island, we were more or less safe from attack from the fly.

Let me describe Mpulungu. The chief thing about the place, I think, was the name; secondly, it is to be noted that there were safe moorings, and that a small jetty had been built, for a boat calls once a month from Kigoma. Then there was a store-shed that Mr Venning had built, so that any produce or supplies that might be brought down to this primitive port could be stored until the arrival of the steamer. It was a simple affair, consisting of four brick walls, about twenty by thirty feet, with a roof of corrugated iron. And that constituted Mpulungu.

However, specially for our visit, there had been constructed, a little way off, an open sun-shelter, which consisted of a thatched roof supported on four standards. The standards were connected by a kind of banister-rail about four feet from the ground, and the space between this rail and the mud floor was filled in with more thatching. A table and chairs completed a very fine alfresco habitation. It was here we discovered that a splendid luncheon had been prepared for us by Smith, whose boys had got busy immediately on our arrival. Sheltered from the tropical sun by our thatched roof, and amid these picturesque surroundings, we had a happy meal.

A runner had been sent to Venning, the Commissioner, with a message telling of our arrival. The Commissioner's residence was at Abercorn, away on top of the hill, 5500 feet up, while we were on the shore of the lake, only 3000 feet above sea-level.

We were just finishing our repast with a cup of tea when suddenly dark clouds loomed up on the horizon. We were in the middle of the rainy season, and on surveying the sky it looked as though we were directly in the path of one of those violent storms that we had been successfully dodging while in the air on our journey down the lake.

Here we realized the helplessness of being on Mother Earth, for there was no escape. Before we could dash from our open sun-shelter to the refuge of the storehouse the rain burst upon us, and so we had to make the best of the shelter afforded by the thatched roof, but as there were no walls to our establishment the driving rain soon gave us a thorough wetting, completely flooding the place.

It so happened that some of the crew had left the luncheon-table early, and were working in the cabin of the flying-boat when the downpour started, and had got a thorough drenching before they could cover up the cockpit of the machine. But, much worse, a tragedy was narrowly averted.

When we left Malta the King's Harbour-master presented my wife with a canary, which he assured her would bring luck, and so the bird had travelled in its cage, hung up in the hull of the Singapore. Daily it was fed and watered and its cage cleaned, and on landing it was usually hung up in the open air beneath the shade of our top wing.

This procedure had happened at Mpulungu, but

in the excitement of putting on the cockpit covers
and closing up the hull against the downpour of rain
the canary was momentarily forgotten. Suddenly
Bonnett remembered that the canary was out in the
rain, climbed out, and made a dash for the cage.
Beneath the wing they discovered the poor little bird
lying on his back nearly panting his last, almost
drowned with the violence of the tropical storm.

However, Bonnett seemed to understand birds, and
by taking him out of the cage and placing him inside
his shirt, next to his warm body, Vittorioso, for that
was his name, soon revived, and furthermore finally
completed the journey to England.

That evening Venning eventually arrived in the
sidecar of a motor-cycle, his car having broken down
on his first attempt to reach us, and by candle-light
we had our evening meal. There was much to talk
about regarding our trip and affairs of the country in
general, and soon the time came to turn in.

Venning was going to sleep in a little tent bivouac,
and the Smiths very kindly provided two camp-beds
for us in the storehouse. Worrall seemed to be very
happy away on some cotton bales in one corner, while
the Smiths said they were going to sleep on board
their motor-boat. Green, Conway, and Bonnett were
going to turn in on board the Singapore.

The arrangements would have been admirable, but
that night it started to rain again, and, what was
worse, it thundered and lightened. Each raindrop
seemed to be about the size of a cupful of water, and
they were falling at the rate of thousands a second

on our tin roof. The roar was terrific, and added to this was the repeated thunder, which resembled high explosives bursting. This terrifying situation was brilliantly lit by a perpetual sheet of lightning, but I think it was the terrible drumming of the downpour of rain on the tin roof that was most nerve-racking.

However, our roof and four walls kept the rain out, and finally it ceased, and soon the dawn came.

Poor Mr Venning had been literally washed away. The Smiths in the motor-boat had been absolutely swamped out, and the remarks of my crew as they came ashore were more than expressive of the kind of night they had spent. We looked a sorry lot as we had our breakfast under the still-standing thatched roof.

During breakfast we discussed our plans for the day. Low clouds appeared to be drifting about everywhere. Our next hop was a short one, a matter of some 200 miles, but on this journey we had to fly eastward over the high plateau and mountains to the northern end of the Nyasa Lake, where we were going to land at Vua.

From where we were at Mpulungu we were more or less 2000 feet down in a trough, for the high plateau over which we had to fly to the Nyasa Lake was anything from 5000 to 6000 feet above sea-level, although the Nyasa was down in another trough, that lake being only 1500 feet above sea-level.

By about noon the clouds lifted, and it looked as though there was going to be a temporary break in the weather, and as most of the bad storms were coming from the west, and our next flight was due east, it looked fairly safe to run through to Vua. By early afternoon we had said good-bye to our kind friends, and were soon racing over the water and climbing away.

The land rises sheer out of the lake on all sides round Mpulungu, therefore, in order to get up above the plateau-level of the surrounding country, we had to start climbing immediately, but even so we did not have to circle, for although the ground rose some 2500 feet between the lake and Abercorn, which was on our course, only ten miles away in a straight line, we flew direct for Abercorn, the Singapore climbing easily, and so rapidly that when we passed over Venning's house we were at least 1000 feet above it.

Then came a country of vast, rolling uplands, covered with deep green vegetation. A few miles away to the north is the frontier of Tanganyika, where the land goes down again to a lower level, and is not nearly so attractive. Up on these highlands it seemed a good place to live, especially as we neared Fife and the head of the Nyasa Lake, for here on these highlands in the near future there will undoubtedly be a White settlement, where the pioneer colonizer may live in peace, comfort, and prosperity, on a fertile land, and surrounded by delightful scenery.

Soon we sighted the Nyasa Lake, walled in at its northern end by an escarpment of brilliantly coloured rock-formations, on which the setting sun caused everchanging hues, and finally we flew above the Songwe

153

Valley down toward the lake until we sighted the small settlement of Karonga, on the north-west shore.

We had originally intended landing at this place, but we understood that it was impossible to get an anchorage there. The Nyasa Lake is like a deep cleft in the earth with straight sides. There are many points along the lake-shore where one can walk only a yard away from the bank and immediately be out of one's depth, and if soundings were taken it would be found that the lake is 300 fathoms deep.

Under such conditions it was impossible to find an anchorage, thus we had been advised to go to Vua, about twenty miles down the lake-shore, where there was a more or less shelving beach offering a good anchorage in three or four fathoms of water for about a mile out into the lake.

Now Vua was marked on the map, but, although we carefully studied our chart, on the first occasion we flew by it without noticing it, and seeing nothing but apparently uninhabited jungle ahead we about turned and flew back along the shore. We remembered having seen two houses buried in the trees some way back, but did not imagine that they could be worthy of mention on the map. On our return flight, however, we noticed a smoke-fire opposite to them on the shore, and so discovered that this was Vua. In all our correspondence prior to the start of the flight we had arranged that a smoke-fire should be lit on the sighting of our flying-boat, so that the smoke-drift would give us the actual direction of the wind near the spot where we wanted to land, for an aircraft must always land and take off head into wind.

No sooner had we got on the water—where, incidentally, a stiff chop was running—than a boat put out from the shore, and by the time we had moored up and our engines had stopped we could hear the cheery voice of one whom we got to know very well later on—Captain Rhoades, late Senior Naval Officer of the Nyasa Lake Marine.

There was a stiff sharp breeze blowing from the north-west which caused quite a rough sea, with the result that it was most difficult to bring the rowing-boat alongside. Refuelling was out of the question, in fact it was most difficult for us to get from the Singapore into the boat at all, and it was very lucky that we had an old sailor like Rhoades in charge. However, we successfully got all our baggage transhipped, then my wife neatly jumped into the boat, and all went well until Worrall took his turn.

He happened to have his hands full carrying one of our self-starting magnetos, which we were taking ashore that night in order to carry out a slight repair. He sat down on the side of the Singapore, and as he went to take his spring into the rowing-boat the leg of his shorts caught on one of the bollards, which left him hooked on the side of the flying-boat. Suddenly there was a ripping and tearing, and, the boat being a yard or so away from the Singapore at that moment, he slipped into the sea. However, he kept his hands above his head, with the result that the magneto was thrown forward into the skiff without getting a wetting or being damaged. I noticed that Conway, who was already in the boat, rescued the magneto and stowed it safely before turning his attention to the man overboard.

It so happened that Dr Child and his nephew were on a visit to Vua in search of a little game-hunting, and they greatly helped in our landing arrangements. As soon as we got ashore a warm welcome awaited us, and we were escorted up to the house. Over dinner that evening we learned of all that Rhoades had done for us. Having taken over the job of putting down our moorings he looked forward to our arrival, but hearing of our delay went north to attend to a little estate he had at Karonga. One day a runner suddenly came through with a message to say that a cable had been received stating that we should be at Vua on the following day. Whereupon Rhoades had 'downed tools' and walked thirty-five miles in twenty-four hours in order to be in time for our arrival. It must be remembered that Vua is a region of no roads, and the only way to get about is on one's feet by native tracks through the jungle.

The dinner-party was a merry one, great humour being provided by Rhoades, who endeavoured to eat with his own teeth some buck that had been shot by the hunters that day. If Rhoades succeeded he was the only one, for all the rest of us failed utterly.

I shall always remember how fascinated my wife was with the whole experience, especially the great attraction which our lamp had for scores of varieties of the things that fly. There were moths and beetles and flies mixed up with mosquitoes and flying ants, but most amusing of all was a certain type of beetle who, after crashing himself up against the lamp-glass, fell on to the tablecloth and started to crawl away. My wife, evidently wanting to stop the progress of one of these creatures, gently placed a small plate on top of his hard back. A few seconds later she was greatly alarmed when she saw the same plate moving quietly across the table, the beetle being so strong that he could carry and drag a weight about fifty times greater than himself.

That night we slept on the veranda, having carefully tucked our mosquito nets round us before going to sleep on the camp-beds. The next morning we were up before dawn as we were going to take part in a little hunt, for here we were right in the middle of the jungle, the virgin bush being only a couple of hundred yards from our doorstep. My wife, Worrall, young Child, and myself made up the party, and it was quite an exhilarating experience as we pushed out into the bush just before dawn hoping to meet great sport. In reality it was possible for us to meet anything from elephant to lion, or from leopard to rhino, but as it turned out we encountered no such dangers, and apart from turning out a couple of jackals, and seeing a few gazelles in the distance, we had no luck.

We decided to continue our journey after lunch to our next halt, Fort Johnston, at the extreme southern end of the Nyasa Lake, and, on hearing how long it was going to take Child to get back to Zomba, I offered him and his nephew a lift as far as Fort Johnston, whereupon they instantly curtailed their hunting expedition by ten days in order to have the opportunity of flying down the lake. This meant leaving all their boys behind to bring their baggage along as best they could.

After lunch we said good-bye to Rhoades, with many a handshake, and were soon flying southward. It was while flying down the Nyasa Lake that we encountered a very strange spectacle. We were flying fairly low, and I could see ahead great pillars of smoke rising vertically out of the lake. There were at least a dozen of these, and after puzzling over the phenomenon for some time I asked Child, who knew the country, what it could be. As soon as he spotted these pillars he said " Kungu fly," and then described to me how the eggs of these flies hatch out of the water. It appears that the hatching of the eggs happens in quick succession, and as they hatch so they fly immediately. The result is that they ascend into the air like a pillar of smoke. We flew right down low as we passed them, and Bonnett was able to get some excellent cinematograph pictures of this phenomenon. Later we discovered that these flies drift toward the edge of the lake, and more or less swarm on to the branches of trees, whereupon the natives instantly seize upon them and collect them in buckets and cans, for they are considered a wonderful delicacy when eaten as we would eat jam or butter.

Farther on down the lake we crossed over to the eastern side, and flew over what is known as the loneliest cathedral in the world, there being a mission station situated on a little island just off the Portuguese shore. It looked a heavenly little spot, and most valuable work has been done there in the past.

As we neared Fort Johnston we crossed the tail end of a storm period on its eastward track, and, successfully dodging the rain, we were able to get some remarkable heavy storm-cloud pictures against the late afternoon sun.

Another wonderful reception awaited us at Fort Johnston, where we were met by Commander George, R.N.R., of the Nyasa Lake Government Service, who had made such excellent arrangements for the accommodation of our flying-boat.

When we got ashore we discovered that scores of folk had journeyed down to the bar where we had landed from Fort Johnston proper, some miles away, and there were many more who had journeyed all the way from Zomba, the capital of Nyasaland, a journey of eighty miles.

As it was essential that I should visit Zomba to discuss with his Excellency the possibilities of the African air route passing through Nyasaland, I found it necessary to leave Worrall and our crew behind in charge of our machine on the lake, while my wife and I journeyed by car overland to Zomba in the mountains. Had there been a convenient lake or river at Zomba that we could have landed on with our flying-boat we would have flown, but this was out of the question.

We spent the first night of our arrival with Mr

L   159

A. C. Kirby, the Resident Commissioner at Fort Johnston, and we could not help noticing the great heat compared with places that we had stopped at farther north. This difference was undoubtedly due to the lower altitude of the place—namely, fifteen hundred feet instead of three or four thousand feet above sea-level.

Nyasaland is a most fertile country and a great tobacco land, and we met many of the growers, who seemed to be troubled with the difficulty of finding a market for their produce.

The journey to Zomba, through delightful scenery, was at the same time most interesting, and my wife had the amusing experience of being almost raided by a batch of monkeys as her car came to a standstill on a hill. They swung in the branches of the over-hanging trees, while dozens walked about on the road in front. All wanted to be most friendly, although I believe familiarity in such cases often breeds contempt. These monkeys are a great trouble to the natives, for they deliberately destroy their crops.

It was about four o'clock when we pulled up on the side of the road outside an assistant District Commissioner's house, and begged a cup of tea. We sat on his veranda, which commanded a magnificent view of a deep valley, and came to the conclusion that his lonely life had its compensations.

At Zomba our host was the Chief Secretary, Lieutenant-Colonel W. B. Davidson Houston, whose house, built under the mountain, had some of the most wonderful panoramic views I have ever seen.

LADY COBHAM WAITING FOR SIR ALAN TO BRING THE CAR ACROSS THE LAGOON

IN A FLYING-BOAT

The Governor, Sir Charles Bowring, entered whole-heartedly into the air-route scheme, and called a meeting of his Government so that I might explain this scheme to all concerned. We also journeyed down to Blantyre so that I might lay the scheme before the Chamber of Commerce, and after four or five days in Nyasaland I felt that I had done all possible to make those in authority quite familiar with the potentialities of the air route. The great difficulty was to decide whether the air line should come down through Nyasaland and so on to Salisbury and thence to Bulawayo, or whether it should go south-westward after leaving Tanganyika, through Northern Rhodesia to the railway line at Ndola or Broken Hill, and then *via* Livingstone to Bulawayo.

There was one incident in Nyasaland we shall never forget—namely, a display of the King's African Rifles. Zomba is the headquarters of this great native regiment, and the Commanding Officer, Lieutenant-Colonel Hawkins, very kindly arranged that Bonnett could take cinematograph pictures of the troops at drill and work, and in the same way Major Stevens, Chief of Police, showed us how the wonderful body of native *gendarmerie* were trained from the raw material.

Reluctantly we had to leave Zomba and motor back to Fort Johnston, and on the way we searched for possible sites for landing-grounds, and made many inspections for this purpose.

Six days after our landing at Fort Johnston we once more took off, and started our 450-mile jump over the land with our flying-boat toward the Portuguese coastline above Beira. Very soon we were flying over Zomba, then came Blantyre, and finally we left the high hinterland of the African continent and came down to the lowlands of the seaboard, when we hit the Zambezi river. Here the weather completely changed, and we discovered that the heavy rains were less local and more difficult to dodge. However, we followed the Zambezi without getting into any storms, although we were very near to them at times, until we came to the mouth of that great river and to the Indian Ocean.

From this point onward we were going to say good-bye to rivers and lakes and content ourselves with following the coastline of the African continent for some twelve or fourteen thousand miles until we reached Tangier, where we should cross over to Europe at Gibraltar.

Chapter VIII

FROM BEIRA TO LAGOS

AS we flew southward toward Beira we continually passed over vast herds of game, and eventually came to the mouth of the Pungwe river, on which the port of Beira is situated.

On circling, we discovered that our moorings were in a very rough sea close to the mouth of the river and exposed to the swell of the open sea. We managed to get down safely, but I instantly decided that it would be dangerous to leave the machine in such an exposed position, to say nothing of the great difficulty of refuelling on such waters. On consultation with the harbour-master I found that there was an emergency mooring much farther up the river in perfectly calm waters, and so I decided to go there. Then came the difficulty of getting off in the rough water, but, worse still, the finding of a clear fairway among the hundreds of small craft, motor-boats, and launches that had come out to greet us.

My wife by this time had boarded the harbour-master's pinnace, and was doing her very utmost in French and English, aided by a whistle, to clear the course. However, I knew that the Singapore would hop off the water quickly, because we were lightly loaded, all the baggage having gone ashore and there being little fuel on board, and again we were in the denser air of sea-level. Added to this there was a stiff breeze blowing, so that when I finally put the nose of the machine head into wind, and pushed the throttles open wide, we leaped forward, and after a run of only a few yards jumped straight into the air. I was told afterward that all the onlookers were thrilled with the demonstration. After turning and flying for a few minutes down the river we landed again and picked up our moorings in calm water under the lee of an island in the middle of the stream.

Our Portuguese friends gave us a hearty reception, and after the festivities were over I gathered the crew together and asked them to draw lots to decide who should sleep on board for the night, as she was so far away from where we were all staying. They would not hear of splitting up, and so Green, Conway, and Bonnett all decided to go down together and sleep on board that night.

The air was balmy and warm. In the sky there were a million stars, and above all there was a full moon, and so Worrall and I said good-bye to them as they pushed off upstream for what they considered was going to be a somewhat interesting night.

I think it was about three o'clock in the morning when I was awakened by a terrific clatter. As I opened my eyes in a sort of dazed condition I discovered that the bedroom was brilliantly lighted, and after a few seconds went into complete darkness again. A moment later there was a crash as though the end of the world had come, and in an instant I realized

SIR ALAN SIGHTING THE INDIAN OCEAN   164

that a terrible thunderstorm was raging. The thunder and lightning continued until dawn, and the rain came down in a deluge the whole time. As the day broke, however, there was a little lull in the heaviness of the downpour, and whereas the visibility had been between fifty and a hundred yards through this terrible deluge, it increased until we could see objects about two hundred and fifty yards away. It continued to rain for hour after hour, and some one said that it would stop by nine o'clock, but when that hour came the deluge was so great that you could not see across the street. It kept on in this manner right through the day, until finally, at about four o'clock in the afternoon, it abated to a drizzle.

All this time our crew had been existing in some form or other on the flying-boat. No one was able to go out to help them, so there they had to remain. Our hull-bottom was undoubtedly waterproof, but it turned out that our hull-roof was not, and furthermore the canvas coverings for the gun-rings and cockpits finally gave way, with the result that Green, Conway, and Bonnett were quietly drenched for some sixteen hours, for it appears that the rain started at about eleven o'clock on the previous night, soon after their arrival on board.

I sent Worrall down to get them in, as I knew there would be voluble remarks regarding the short-comings of flying through Africa, and no doubt there would be some very pointed comments upon the subject of sleeping on board for the night. I felt it far better that we should discuss the next day's

165

journey after they had all had a bath and dry clothes, a good square meal, and a bottle of wine.

On the following day the sky, seeming to have run itself dry, was all blue and the earth was flooded with sunshine, and so we said good-bye to our new friends, and, pulling ourselves off from the waters of the Pungwe river, we followed the coastline of Africa southward toward Lourenço Marques.

The Indian Ocean offered little refuge for our flying-boat, as it would not have been practical to have attempted alighting on that sea, owing to the giant rollers that are perpetually thundering on to the beach. The long, deep swell is continuous in this part of the world. However, the coastline was ideal as a flying-boat route, as a string of lagoons ran parallel to the sea all along this shore, and there were also numerous river mouths.

After we had been going a couple of hours Green came forward and intimated that he wanted to speak to me, so nodding to Worrall I left him at the controls and went down below. Green took me to the instrument-board and showed me the variation in the water-temperatures of our two engines. One was excessively high, and Green seemed to be worried about it. I asked him if he would like to have a look at the engine, and he replied, " Of course, but there is not much chance of it." Whereupon I told him that I could land any moment he liked, and drew his attention to the lagoons that we were passing over.

I was soon back in the cockpit and could see just what we wanted a few miles ahead. It was a fine, long lagoon lying in a valley bordered by the rising mainland on one side and the sand-dune that separated it from the sea on the other. A stiff wind was blowing in from the sea, which meant that I should have to land across the lagoon, the narrow way of it, but even so there was plenty of room. As we approached there appeared to be plenty of depth of water, and, having already told Worrall of my intention, I pulled the throttles back, and in a half-circle glided in to land. As we came down to the level of the lake we were dead into wind, and while I was holding the machine off, and looking into the water ahead of me, to my surprise I could see the clean, sandy bottom with white pebbles on it, through the transparent water, as clearly as if there had been no water at all. For a second or so I had a shock, for I thought that I was about to land in a foot of water, which would have been far too shallow for us to come down on. While I was still holding the machine off, however, I realized that it was much deeper than I had first estimated. The next instant we touched, and with the noise of rushing water on our hull we glided on to the surface of the lagoon. Presently all was silent, and we were floating on the water. Worrall, who by this time was up in the nose of the machine making preparations to drop our anchor, motioned to me to turn the switches off, which I did, bringing our propellers to a standstill.

Then I climbed forward to give Worrall a hand with the anchor, and together we dropped it overboard, watching it sink into about twenty feet of

167

water, but so clear was this lagoon that we could see it on the bottom on the white sand. As we drifted backward the breeze took charge of our big craft while we paid out our anchor-line until such time as the angle of the rope would give the flux a good chance to dig deep into the sand. Then with a couple of turns over the bollard we made fast.

The anchor held, and we set to work to attend to the business of the high water-temperature, but Green was already on the job, and had discovered a small leak in our extra tropical radiator, so that since the start of the flight we had been losing water.

The leak was not a serious one, and as it would be a long job to effect a repair I decided that it would be quite safe just to fill up the radiator again and continue the flight, being confident that we could reach Lourenço Marques before more water was required.

The trouble was where we should get the water from. My wife immediately asked, " Why not out of the lagoon? " Whereupon somebody said it would be salt, so in order to settle the question Conway leaned out of the rear cockpit and scooped up a glass-ful, handing it to me. Every one silently waited for the verdict, for if it were salt we should be in an awkward predicament, being a score of miles from habitation or place where we might get fresh water. However, I tried it and found it to be quite fresh, although a bit flat. We lost no time in filling up the radiator, and then drew up our anchor. While the rope was being coiled we drifted back toward the shore behind us in order to give us the maximum run to take off across the lagoon. It was not until the two engines were running that we finally put our anchor away, and opening out we were soon in the air once more, with our water-temperature down to normal, heading for Lourenço Marques.

When we had arrived over this wonderful seaport of Portuguese East Africa it was simplicity itself to land on the sheltered waters. We were met by the British Consul-General, Mr Pyke, who had taken the whole matter of our visit so capably in hand. The Portuguese harbour-master had arranged every facility for us, and as soon as we had made fast the fuel was ready to come on board; so leaving Worrall in charge of the machine, its final mooring, inspection, and refilling, my wife and I went ashore with the baggage in a waiting pinnace to attend to all the necessary social duties.

Mr Pyke was a most wonderful host, so that when we came to leave, after a full day's stay, I could not help feeling that if every visit would go as simply, smoothly, and take place under such ideal conditions, the successful completion of our flight would be a very easy matter indeed.

When we flew on to Durban we took Mr Pyke along with us as passenger, and I believe that the glorious panoramic view down the coast to Natal gave him a great thrill.

We had warned Durban of the time of our arrival, and I was determined to alight to the nearest minute, because it was necessary for the harbour to be cleared, and for the shipping to be more or less under control, if we were to land in safety. As soon as we flew over the town crowds seemed to rush to the harbour beach, and I heard afterward that many girls spoilt their shoes and silk stockings with running over the sands, for the tide was out. Finally I got into a glide, in order to land up one of the prescribed channels, according to the directions on the chart that had been sent by the port captain to await my arrival at Lourenço Marques. All my final instructions about each particular port, throughout the flight, we invariably picked up at the previous port of call, having been posted back to us by the harbour-master or port captain.

Our landing at Durban was possibly the most thrilling of the whole flight. To start with, we were the first people to fly from England to Durban, and, secondly, our craft was the first flying-boat that South Africa had ever seen, and Durban seemed proud of the fact that we had landed in her harbour, for as we touched the water thousands along the quayside burst into cheers.

Our reception at Durban was wonderful, and the kindness shown to us by every one made us reluctant to continue our journey. The harbour authorities placed everything at our disposal, for this was to be our main depot for overhaul before the homeward trip up the west coast. We intimated that we should like to inspect our hull, whereupon the port captain had us towed up to the new dock, and the latest electric crane, with absolute ease, lifted our flying-boat out

LADY COBHAM IN TOWN ATTIRE AT DURBAN   170

of the water and swung her over some sidings into a sheltered position from the prevailing wind, behind a building. While the flying-boat still hung in the air we lashed her to the railway lines to keep her rigid.

We met so many kind friends in Durban that it would take a book to mention them all, but Mr and Mrs Leuchars seemed to take us under their wing from the start, and they did all they could to make our stay in Durban a happy one, while Mr Bob Wright, that well-known personality of the Durban port, seemed to be always at our disposal, eager to do anything he could to help.

It was at this stage that I left Worrall in charge of operations, with instructions to carry out a complete inspection of the machine, while Green was free to get down to work on his engines. I then made a journey by motor-car from Durban to Johannesburg, as I wanted to study that particular piece of air route from the ground.

The first sixty miles of the journey by way of Maritzburg rises in a series of mountain ranges to an altitude of between five and six thousand feet, and on to the general high plateau of the hinterland of South Africa, where the weather conditions are totally different from those prevailing on the coast. On the road up we ran into fog, low cloud, and rain, and I saw the weather at its worst, which is a very good thing. Then at Johannesburg, where I was the guest of Dr Samuel Evans, that great aviation enthusiast, the Johannesburg Aero Club very kindly lent me their Moth, and their chief instructor as companion, so that I might make a tour up into Southern and Northern Rhodesia.

I wanted to confer with the Governor and the Government of Southern Rhodesia regarding the possibilities of the through-Africa air-route scheme, and I wanted to do the same with the Governor of Northern Rhodesia, and if possible to enlist the support of both territories. At the same time it was necessary to come to some decision as to the correct route for the air line to take after leaving Southern Tanganyika.

We left Johannesburg at about 6 A.M., and flew 250 miles to Palapye Road, where we had some breakfast, and then, after a similar journey, flew on to Bulawayo in time for lunch. The afternoon was spent in a series of interviews with various business folk, and the following morning we left before 7 A.M. on another 200-mile journey to Salisbury, the capital of Southern Rhodesia, where I had an appointment with the Governor at 10 A.M. I had another at noon with the Premier, and after a full day's business flew away early the next morning in order to keep a breakfast appointment in Bulawayo with my old friend Mr Barber.

Everything went according to plan, and at 10 A.M. I again took off from Bulawayo, this time to keep a luncheon appointment with his Excellency in Livingstone, the capital of Northern Rhodesia. I followed the course of the railway line through the trackless bush until I came to the Wankie Collieries, when in

PASSING OVER THE VICTORIA FALLS

the distance on the horizon, in the direction of my destination, I could see what appeared to be the rising smoke from a little bonfire. But I had been out over this particular country before, and I knew that what appeared to be smoke on the horizon was eighty miles away, and was nothing more or less than the spray from the Victoria Falls rising 1000 feet into the air. And so for an hour I flew toward this object, until finally I was flying over the wonderful spectacle of the mighty Zambezi, which is a mile and a quarter wide at this point, falling 400 feet into what appears to be a crack in the earth.

All this avalanche of water escapes out of one narrow channel, which zigzags its way out through the plateau to lower levels.

After a busy day in Livingstone I returned on the following day to Bulawayo, and on the next was back in Johannesburg once more, having accomplished in a matter of a few days a tour that would have taken at least a fortnight by the existing modes of transport. Furthermore I had been successful in securing the support of both Northern and Southern Rhodesia.

Before leaving Johannesburg I journeyed over to Pretoria to visit Sir Pierrie Van Ryneveld, Commander-in-Chief of the South African Air Force, and undoubtedly the greatest exponent of flying in South Africa.

Back in Durban once more we were soon ready to continue our journey, and so, having said good-bye to all the dear friends who had helped us so much, on a wet and drizzly morning we took off out of Durban

Harbour. On board I had my old friend Mr Merckel, my London lawyer, who happened to be on a holiday cruise to South Africa. We were going to fly him to Cape Town in order that he might overtake the boat that would carry him back to England.

The journey from Durban to Knysna was a matter of 600 miles, and considering that it was the first flight that Merckel had ever made he was in for a good long one. We took off in drizzle, and later passed into storm, and at Port Elizabeth we skimmed over the town in fog and rain. It was not until we got within forty miles of Knysna that the weather cleared and the sun came forth. It was one of the worst flights of the whole trip, and the continual downpour of the heavy rain would, I knew, play havoc with our propellers. For many miles on the trip the visibility was only 400 yards, and, as Worrall remarked, it was "just like flying at home."

We landed in sunshine in the picturesque little port of Knysna, which is shut off from the sea behind two mountain walls, there being a narrow gorge which serves as an entrance to this secluded anchorage.

Amid scenery that reminded one of Norway or Sweden we enjoyed charming hospitality for the night, and on the following day did another 500-mile flight to the Cape peninsula.

We approached the rocky mountain prominence from the east, and as the peninsula came into view we discovered that its mountain peaks were covered with a blanket of billowy white clouds, with here and there a jagged rock-spur standing up out of the sea of mist all round. Soon we were passing above it all, and as we looked down on to this snow-white carpet below us, occasionally there would be an opening and we would get a glimpse of the famous vineyards of the Cape, thousands of feet below, or the next instant a jagged mountain peak only a few hundred feet beneath us, and then finally we could see Cape Town itself.

Bonnett was busy taking pictures all the time, until I decided that we must come down and land, and finding a convenient opening in the clouds we passed through them, and, once below the layer, the whole vista of Cape Town, Muizenberg, and Simon's Bay, our landing-place, lay before us.

A warm greeting awaited us from the Mayor of Simon's Town, after which we were raced all the way to Cape Town, in cars of British make, to be the guests of the Lord Mayor of that city on the occasion of a luncheon that he was giving in the combined honour of the British Air Force Aeroplane Flight and ourselves. The former, having completed the first stage of their flight from Egypt to South Africa, were returning northward on the following day.

We all felt that from this point we were turning the nose of our machine homeward, and that no more should we be flying farther away from England. On the contrary, every hour that we flew would bring us about a hundred miles nearer to Rochester, our final destination and our starting-place.

On the particular morning of our start from Simon's Bay a stiff south-easterly breeze was blowing, which rendered the sea so rough that it was impossible to get alongside the rear of the hull of the Singapore without running the risk of damaging our craft. To row against the breeze was a physical impossibility, and finally we got on board with the aid of a tug, for while it steamed head into the gale from its stern they lowered us down in a skiff on to the nose of the Singapore, and we all managed somehow to clamber on board from this position.

All this time we were in the lee of a huge break-water, so that the sea was very short. Having started our engines we cut adrift, and with our propellers ticking over we drifted backward toward the shore in order to get sufficient run to enable us to take off without going into the open sea. When we were within fifty yards of the shore, our engines being warmed up, we opened out, and were soon racing headlong for the great jetty, but after a run of a few hundred yards, head into the gale, we had flying speed, and pulling our controls back we leaped off the water and climbed away, clearing the breakwater by several hundred feet.

At about 500 feet we turned the nose of our machine northward, and were soon flying toward Cape Town with a 40-mile-an-hour wind behind us. I estimated that we should reach our destination, Lüderitz, 500 miles away, in record time, but to our surprise we discovered that five or six miles beyond Cape Town the wind began to drop, and ten miles north of the peninsula we were flying in a more or less dead-calm atmosphere, with a flat, oily sea beneath us, which all goes to demonstrate the erratic weather conditions in this part of the world.

When we had started it had been a cold, chilly morning, but by the time we had reached Saldanha Bay, that magnificent sheltered harbour sixty miles north of Cape Town, the atmosphere had become quite mild, and when we had left Saldanha fifty miles behind us, although we were flying at 3000 feet, the air had become excessively hot. The farther north we flew the hotter it became. Green began to look anxiously at the oil-temperature gauges, and a little more than 100 miles north of Cape Town, when Table Mountain could still be seen on the distant horizon behind us, our water- and oil-temperatures were registering higher than on our whole trip through the African continent. This was a most extraordinary state of affairs, for not even in the Southern Sudan had the atmosphere in the cabin of the Singapore been so stiflingly hot.

The next hour's flying was an anxious time for us, for we did not know when the temperature was going to stop rising, but, fortunately, near Port Nolloth it became constant.

On this trip we passed the mouth of the famous Orange River, where the new diamond fields have been discovered recently. I remember being advised by a mining engineer, while in Canada some two years before, that if ever I flew out to Africa again I was to take him with me in a seaplane or flying-boat, whereupon we would go secretly to the mouth of the Orange River. My informant impressed upon me that all we should need to do would be to land on the waters of the Orange River just above its mouth, drop anchor, and there spend a quiet week-end with a bucket and rope, and he was convinced that we could haul out of the bed of the Orange River more diamonds than our flying-machine would be able to lift out of the water. Allowing for exaggeration on his part I had always cherished a little idea in the back of my mind that here was the golden opportunity for our little expedition to " get rich quick," thinking that something might be done on the lines suggested by my friend in Canada.

I had visualized the possibility of my crew and myself spending a few days in shovelling up diamonds from the bed of the Orange River, and flying on with our secret hoard of treasure. However, there was no such luck, for, although the diamonds were undoubtedly there, we were too late. They had already been discovered, and the Government had more or less taken control, so that there were armed guards posted all round the district ready to shoot on sight.

Contenting ourselves perforce with the more peaceful occupation of air-route development, we continued our journey northward to Lüderitz, where we landed in the sheltered harbour soon after midday.

Lüderitz is famous for two things—crayfish, which is caught, canned, and exported, and diamonds, which are retrieved from the sandy wastes by a gigantic washing-plant. In this town there is no water-supply whatsoever—in fact, water costs up to sixpence a gallon, as the entire supply has to be distilled from sea-water. However, we discovered at the splendid reception that they gave us that they had better things to drink in Lüderitz than distilled water.

After a whole day there we reluctantly left our kind friends and set out on one of the most treacherous journeys of the whole trip—namely, the short flight of 250 miles up this fog-bound coastline to Walvis Bay. It so happens that the permanent wind in this direction is south-westerly, coming direct from the South Pole over the cold waters of the Antarctic Ocean. These winds are full of moisture, but they are also very cold. Consequently, instead of coming down in rain when they strike the hot desert coastline of South-West Africa, they go up in steam, causing a perpetual fog.

The whole coastline and district is more or less unsurveyed, and very little is known about it, especially from the flying point of view. Fearing that our destination, Walvis Bay, might be buried in fog, we could not run the risk of climbing above the mist bank, because we might not be able to see Walvis Bay when we arrived over the site, and if the fog were low we should be unable to find our way down on to the water.

For hundreds of miles inland there is nothing but absolutely treeless, trackless, waterless desert—a veritable sea of rolling sand-dunes. Therefore, under these conditions, our only course was to endeavour to hug the coastline and fly under the fog.

As we continued on our journey we discovered that heavy fog-banks rolled in off the sea, and as though in an endeavour to get up on to the high sand-dune cliffs they jumped from the breakers to the top of the sand-dunes and then rolled on inland. Thus there was a kind of three-sided tunnel that was partially free from the dense fog. The sides of this triangular tunnel were formed by the beach, the sand-dune cliff, and the curtain of fog which extended from the sea to the top of the cliff, and down this treacherous pathway we had to fly.

I was only too glad when, after nearly three hours of flying, we came to Walvis Bay and found the harbour fortunately clear. We lost no time in landing on the glassy surface of the bay, having completed a very nasty stage of our journey.

We discovered that the wonderful natural harbour of Walvis Bay was growing rapidly from a place with a name and no inhabitants to a thriving port with an increasing population. The whole town turned out to meet us, but, like all other places at this stage of our flight, we had to say good-bye the following day.

Our next port of call was to be a place called Porto Alexander, a small fishing village situated inside a wonderful natural deep-water harbour. We had decided to call at this small place because we knew that we should be sure of a safe anchorage in its sheltered waters, whereas at Mossamedes, the chief town of this territory, a few miles farther north, the anchorage might have been open to the main swell. The journey up from Walvis was along a continual desert coastline, past the region of Great Fish Bay, formed by a huge sand-spit running out into the sea.

IN A FLYING-BOAT

The story goes that Great Fish Bay is so full of fish that it is dangerous for fishermen to cast their nets there because the weight of the catch would carry them away. However, this seldom occurs, as the coastline is practically uninhabited, with the exception of some dozen natives who live on the end of this sand-spit.

We flew very low along this coastline, often to inspect wrecks, for they were the only relief to the endless hundreds of miles of sand-dunes. When we were about 250 miles north of Walvis Bay, and were flying at an altitude of about 100 feet, we noticed on this uninhabited, lifeless beach a wolfish-looking dog scampering along. When we arrived at our destination we discovered that wild dogs are known to inhabit this coastline, which is all the more strange considering that there is no fresh water whatsoever for hundreds of miles inland, nor any vestige of vegetation. It is surmised that these dogs must live on fish that they grab out of the sea, and furthermore that they must drink salt water.

The only visible sign of life on the whole length of this coastline was afforded by the flocks of birds skimming low over the water. We were passing over them continually in their millions, so that they practically blotted out the sea beneath us. I was terrified lest they might rise suddenly and dash into the propellers and wings of our flying-boat. I doubt whether they would seriously damage the structure of our machine, but if they flew into our propellers they would certainly cause them to fly into a thousand pieces.

At last, on the horizon, we saw what was to us at that moment a strange sight—namely, two or three trees, for this was the first sign of any form of vegetation whatsoever for nearly 2000 miles. It was at Porto Alexander, and soon we were flying over what we had expected to see—namely, a sheltered harbour made by a sand-spit spur running parallel to the coast. Below we could see a gunboat at anchor in the middle of the bay, and, according to my chart, I made out the moorings that had been specially laid down for us.

No sooner had we tied up and made fast than a pinnace came alongside from the gunboat, and we were escorted to that ship forthwith. I then discovered that the Portuguese Government, on hearing of our determination to go to Porto Alexander in preference to Mossamedes, the big town farther north, had fallen in with our wishes in this respect, but, feeling that Porto Alexander would be unable to give us all the facilities required, owing to the fact that it was such a small, isolated village, they had sent a gunboat some 600 miles from Loanda in order to await our arrival and give us assistance. I looked upon it as the greatest piece of hospitality and courtesy that could have been bestowed upon us.

The Captain had prepared a sumptuous luncheon for us beneath an awning on the upper deck. Thus sheltered from the sun we sat down at a long table. A gentle breeze was blowing, and the whole situation was delightful. We started with an *apéritif*, and then followed course after course until some dozen had passed by, and with each one there was a different wine, until eventually came champagne. Finally, although it was 90 degrees in the shade at the time, this great feast was capped with glasses of port—the wine of our hosts' homeland.

Though we had started the luncheon soon after midday it was now nearly four o'clock, and we were informed that we had to attend an afternoon reception ashore given by one of the village clubs. Porto Alexander boasted a population of some three to four hundred only, but I was informed by the local wag, who spoke fair English, that the village had two clubs and five sets. By four-thirty we were ashore, had made arrangements for the refuelling, and were walking over the sand-dunes to the club-house, where some of us, possibly as a result of our luncheon, were already looking forward to cups of tea. We soon discovered that tea was not the order of the day, but instead there were many toasts to be responded to in sweet champagne. By five-thirty we were walking across more sand-dunes to the other club, where we discovered that there were many more toasts to be responded to in more sweet champagne, and by six-thirty we were struggling along the sandy beach toward the Administrator's house, where we were going to stay the night. There appeared to be a little hurry astir, and I discovered that the Administrator wanted us to have time to attend to our toilet, in preparation for the great banquet that he was giving in our honour at seven-thirty that night.

By this time I began to feel nervous, but I simply

had to carry on, for I discovered that our hosts had been preparing for our visit for weeks and weeks.

Many folk from Mossamedes had travelled scores of miles over the hot, sandy desert in order to greet us, and no doubt many were eagerly awaiting the Administrator's banquet, but they had not been invited to the Captain's luncheon, and they had not been present at the receptions given by the two clubs.

As the dinner progressed I lost count of the courses, and the wines were too numerous to mention. All this was happening at a little fishing village on the edge of the desert, whose livelihood was derived from catching fish, drying it in the sun, and exporting it in little sailing-ships to supply the native market hundreds of miles up the coastline.

The only shadow that overcast the day's events was our engineer's mysterious illness. Ever since we had left Lüderitz Green had been suffering from internal pains, and being an old malarial patient he had had several bouts of fever. No doctor that we had come across at Walvis or Alexander could diagnose what was the matter with him, and Green just lay in great pain in one of the bunks in the flying-boat while we were travelling, and on a camp-bed when we could get him ashore.

The following morning we had to say good-bye to our generous hosts to continue our journey northward.

The coastline, although still desert, had changed from sand-dunes to rocky prominences. We passed over the town of Mossamedes, and then came a series of mountain-cliffs that dropped down sheer into the sea. Here and there were little dark, deep-shadowed bays where my chart told me that an anchorage was impossible, for against the sea-wall itself there were depths marked of ninety and a hundred fathoms. Finally, the lifeless rocks gave way to forms of vegetation, which, with the exception of our few trees at Porto Alexander, were the first that we had seen since passing Saldanha Bay, a few miles north of Cape Town, which was over 2000 miles away.

We were making for Lobito Bay, and in due course what is undoubtedly one of the world's finest harbours could be seen on the horizon. Lobito Bay is the terminus of the Benguella Railway, which is going to bring all the copper ore out of Northern Rhodesia to the coastline. Just before reaching Lobito we passed the old-time port of Benguella, whose prosperous days dated back to the time of the Arab slave-dealer, who marched his captives through bush, jungle, and desert, in the heat of the African sun, toward the port of Benguella.

The road to Benguella was worn level by the feet of human slaves, and the mortality there is an appalling blot on the history of Africa. It was no doubt this that brought forth from Livingstone, after his experiences there, the statement that Benguella was the " open sore of Africa."

The Portuguese Governor's reception of our expedition on its arrival at Lobito was one of the great compliments paid to us on our tour round Africa, and while at this delightful port we received also the

185

kind hospitality of Major Greenwood, Sir Robert Williams' representative in that part of the world.

At Lobito Green had a very bad turn, and lay in a terrible fever all night. We all surmised that he was suffering from a severe attack of malaria, as he had been subject to this complaint for many years. He got somewhat better the following morning, however, and determined to complete the trip, so that we still had him with us when we took off from Lobito and headed for our next port of call, Banana Creek, at the mouth of the river Congo.

We were now nearing the rainy areas and approaching a country of vegetation once more, although the coastline for the most part of this particular journey was a mass of rocky hills, with gaunt cliffs of hard, red stone dropping sheer into the sea. I had Admiralty charts of many parts of this coast, and I discovered that in many of the little silent, lonely bays, with their high rock-walls towering on three sides, the sea below was as much as a hundred fathoms deep within a few yards of the rock-face. For scores of miles an anchorage would have been impossible.

It was Easter Sunday, April 8, and the rainy belt was gradually moving northward, but we expected to overtake it any day, and were prepared to run into tropical storms at any moment. However, we were fortunate enough to reach Banana Creek without encountering rain.

We flew across the mighty Congo, which is about five miles wide for about 100 miles up from its mouth. Banana Creek was a fine sheltered water on the north bank of the Congo, immediately at its mouth, so that it was only a spit of land some 200 yards wide and two or three miles long that separated the creek from the open Atlantic.

The Belgian Government had very kindly sent one of their steamers down from Boma, the capital of the Congo colony, fifty miles away up the river. We all looked upon this as a very great compliment indeed. On board we found the British Consul, who had come down to meet us, and, what was most thoughtful on the part of our Belgian friends, a doctor.

Very shortly after our landing we managed to get Green, whom we had been carrying for many a day on one of the bunks in our cabin, to the Belgian steamer. Here the doctor made an examination, and a few moments later informed me that it was impossible for Green to continue the flight, and that he must take him with all speed back to the hospital at Boma, as he surmised that an operation might be necessary.

The air was hot and sultry, and by sundown Worrall and I, with the aid of some natives, had re-fuelled. Conway had gone through his schedules with both the engines, and, having taken over Green's responsibilities, he informed me that all was in order for the start on the following morning.

That night we slept on board the steamer, and there did not seem to be a breath of air moving anywhere. Poor old Green had a terrible time, but worst of all he was so depressed at the thought of saying good-bye to us all, after having done over two-thirds

187

of the journey, and being, as we estimated, only a week or ten days from home. It was his ambition to complete the trip, and when we said good-bye to him just before dawn on the following morning he broke down completely.

I think the hardest moment of all for him was when he heard the roar of our engines as we took off and flew low over the steamer, which was already under way going full speed ahead for Boma and the hospital.

The exact sequence of events regarding Green we did not hear until many weeks later, but this is what happened. Immediately on arrival at Boma he was taken to the hospital, where the house surgeon decided that Green's complaint was not fever, but internal trouble that needed an operation immediately. Green was put under an anæsthetic, but the operation proved to be of no avail, and the doctors decided that the only thing to save him would be to get the aid of a certain specialist who happened to be over 300 miles away up the river. It was a matter of hours as far as Green's life was concerned, and the problem was how to get the specialist to Boma in time. The difficulty was solved by the fact that the Belgian Congo in the year 1928 possessed an air route, and furthermore the whole country was linked up by wireless. So it came about that the authorities wirelessed instructions that the specialist should proceed in a plane immediately to Boma, and it was thus that this doctor was at Green's bedside in less than four hours after this decision had been made. He operated within five minutes of his arrival, and afterward it was said that

188

in another twenty minutes it would have been too late. The doctor's journey to Green's aid, two years before, would have taken as many weeks by canoe—the old method of transport. We shall always feel eternally thankful to our Belgian friends for their kindness, forethought, and generosity, for without their prompt assistance Green's life could not have been saved. This was a great demonstration of what modern science can do. It proved the practicability of flying and of wireless. The vast impenetrable jungles of the African continent that have baffled mankind for hundreds of centuries will in two or three years be completely conquered by man, aided by the new sciences of wireless and flying. Darkest Africa will be a thing of the past, and this great continent will be opened up and developed a thousand times faster than would be possible otherwise.

It was on the flight between Banana Creek and Libreville that we ran into the rains. Between the storms there was no mist; in fact, the visibility was exceedingly good; but we discovered that the storms stretched over vast areas and were difficult to get round. The rain was very heavy and seemed to fall in enormous spots, but what eventually worried us was the hail. I have heard of hailstorms in these regions in which stones fell as large as eggs. If we flew into such a storm we knew it would mean that our propellers would certainly fly into a thousand pieces.

I can well remember one storm that we overtook. As it seemed to come from the south-east, off the land, our only course to avoid it appeared to be by flying out to sea and getting ahead of it. Our correct course for Libreville was more or less due north, so that when we headed seaward in a north-westerly direction we were still making headway. The trouble was that the storm kept pushing us farther and farther out to sea. To have attempted to fly through it would have been disastrous, for the hail looked like a mass of white balls falling, and on one occasion when we got very near to the edge of this downpour we could see that the hailstones were at the very least half an inch in diameter.

Finally we got ahead of the storm, and, turning our nose north-eastward, we eventually converged with the coast once more, only to run into another storm which stretched from horizon to horizon, and was impossible to get round. I knew that we should have to take refuge somewhere, and on consulting my chart discovered that there should be some large lagoons a few miles to the north, and as there was no refuge behind us the question which presented itself was, " Could we reach our lagoon before we got into the depth of the storm ahead ? "

To land on the open sea was impossible, owing to the permanent rollers, so we ploughed ahead toward a horizon that was blue-black, being continually lit up with forked lightning. It was a great relief when we sighted the lagoon, for the mere fact that we could see it told us that the storm had not broken there yet, and so we raced on until we came to these great open lakes, the far ends of which were blotted out by the downpour. We knew that within two or three minutes the spot where we were flying would be in a similar condition.

We had no idea of the depth of the lagoon, but just had to chance that there were at least four feet of water for our hull to clear the bottom safely. We circled once and landed down an open fairway, and as soon as we came to a standstill we found that we were in about five feet of water, so we lost no time in dropping anchor.

I was worried lest, owing to the force of the gale, our anchor would be insufficient to hold us, but when the rain did come it was not quite so bad as we had expected. It was indeed comparable to the worst downpours we ever get in England, but on investigating the horizon from the front cockpit I came to the conclusion that we were luckily on the edge of the great storm that was passing by.

After half an hour it settled down to a quiet, steady rain, and my fears regarding the gale being over, I decided that we should have to be content to wait, for away in the direction of our destination it looked as though a perfect inferno were raging.

My wife in the meantime cheered us all by preparing lunch, which on this particular day almost resembled a feast. We started with hot soup, our Primus stove coming into action here. Next came crayfish, for we had been presented with many tins of this by the folk at Lüderitz, where, next to diamonds, it is the chief industry. Then came a chicken (cold, of course), but with it we had some bread, or two or three kinds of dry biscuits to choose from. For those who wanted it there was Fray Bentos corned beef, followed by cheese and biscuits. Lastly came fruit, for we always had a goodly supply of oranges and apples on board. This repast was brought to a close when my wife poured out cups of steaming hot coffee, while I apportioned a tot of Van der Hum liqueur to each.

The main thing was to keep up our spirits. All the time the rain was pelting down, the sky was almost as dark as night, and we could hardly see the shores of the lagoon. There we were marooned under these dismal conditions on an unexplored coast, hundreds of miles from the nearest form of habitation. The hours passed by, and my wife did her best with her ukulele to keep us merry. She sang ditty after ditty, many of which were truly funny, and it was not long before she had the crew joining in every chorus.

It must have been nearly four hours after we landed before the rain finally stopped and the sky ahead seemed to clear a little, but instead of the usual wonderful visibility it was like a misty, dirty day at home.

In view of the fact that from this spot right to our destination there were many similar lagoons marked on the map, Worrall and I reckoned it would be fairly safe to continue.

Very soon we had taken off our covers and started up our engines, but to our dismay we discovered that we did not go forward, for our hull was touching the bottom. Looking round I guessed what had happened.

Our lagoon at some point was connected with the sea, and the tide had gone out and taken nearly two feet of the lagoon water with it. We took soundings, and discovered that we were only in about two feet six inches of water, but as the bottom was of soft mud we estimated that we could just about get off, for we knew that when we opened our engines out, provided we could get over the first twenty feet of our run, our hull automatically would rise sufficiently on the water for us to clear the bottom. There was no wind, and so, pointing our nose toward what appeared to be the best fairway, Worrall opened the engines full out. We stuck for a second, then slipped forward on the mud-bottom, whereupon I pulled the controls hard back. Her nose lifted, our hull rose on the water, and in the next instant we were skimming over the surface of the lagoon and gathering speed. Soon we were on the step of our hull, and in some thirty seconds from the start we were off and heading our machine northward toward Libreville. The farther we went the better the weather became. When we finally reached our destination, on an estuary, we discovered that there was a strong wind blowing, and instead of moderately smooth waters, as we anticipated, we found that we should have to land on a very rough sea.

We sighted our moorings according to our forward instructions, and came down and landed. We were unlucky, for just as we touched we hit a rising swell which bounced us, and in a series of thuds we came to rest on perhaps the roughest water we had landed on on the whole trip. However, it seemed to make little difference to the hull of the Singapore, and she took it well. Our greatest difficulty was when we came to pick up our mooring, which we discovered was a conical buoy, about ten feet high. The danger was lest we should overshoot and run the risk of getting the buoy up against our wings or propellers before we had lost way, for this iron menace bobbing up and down in such a terrible sea would certainly have damaged us severely.

All went well, however, and we were soon tied up and drifting well astern of the buoy. At Libreville we found a most energetic British Consul, who arranged everything very nicely for our comfort. After we had all bathed and changed we found invitations awaiting us from the Governor, to attend a *soirée*, given at the Residency especially in our honour. The hospitality showered upon us by our French friends at Libreville we shall never forget. It was about six o'clock, and, knowing the love of all Britishers for tea, they had prepared some specially for us, but of course this was only preliminary to all the good things that followed, and the champagne flowed freely when the toasting began.

When we left, on the following day, presents were showered upon us by the residents, who individually gave us priceless souvenirs in ivory and silk, and somewhat reluctantly we said good-bye. It was early in the morning, and to our joy we discovered that the estuary on whose waves we had landed the day before was now like a millpond, and so without difficulty we

194

NATIVE CANOES GREET US AT BONNY

195

got into the air once more, and this time headed north-west direct for Bonny, at the mouth of the Niger.

On the map it will be noticed that there is here a right-angled corner in the African coastline, and in that angle, a hundred miles out to sea, is the great jungle-island of Fernando Po. On our trip to Bonny the southern shore of the island was going to be our one and only landmark when we cut the corner and made the direct flight across the open sea.

For this occasion we had fair weather, with a gentle wind on our starboard beam, except when we neared Fernando Po. Then we discovered there were strong breezes blowing which were purely local.

The coast of Nigeria is low and flat, and we did not see it until we were within a few miles of it, but our compass brought us up accurately to the mouth of the Bonny river. We landed a few miles away at the old, now obsolete, port of Bonny.

No sooner did we get on the water than I noticed in the distance some enormous native canoes, manned by scores of men, paddling toward us at a great speed, and by the time we had picked up our moorings they were circling round us. Each canoe was manned by about forty men, whose strokes were kept in rhythm by the beat of a drum. In addition to the drummer there was a skipper, who seemed to urge them on to greater efforts, while the coxswain at the stern struggled with a giant paddle to steer the craft. The various canoes presented a wonderful spectacle, and I pressed Bonnett to get into action with his camera as quickly as possible.

While Bonnett was filming them at different angles I was endeavouring to get snapshots of some of the more gaily bedecked canoes, shouting to him directions regarding the various shots. Suddenly I discovered that I was talking to myself, for when I looked round, instead of finding all my crew on top of the hull, as they had been a few minutes before, to my surprise there was not a soul to be seen.

I ran to the rear hatchway to see why they had all gone below, when I was met by a bilge-hose being thrust out. This was followed by Conway, pump in hand, who set to, working the pump for all he was worth.

Without any suspicions I asked him what he was doing with the bilge-pump, for it was one of those things that we had never used since the start of the flight. As he did not reply I casually asked, " Surely we have not got a leak? " Whereupon he replied, " Oh, no! I am only trying to see if it will work, sir," and at that moment water came gushing out of the hose, and I discovered that it was a case of " All hands to," for we had really sprung a leak.

It had recently been discovered that the finest stuff to stop small leaks in metal flying-boat hulls, in a case of emergency, was well-masticated chewing-gum. This particular commodity seems to have the right ' stickasity ' for filling up a small crack or hole and for withstanding the pressure of water. Hence we had taken on board huge supplies when we had left England, but when I rushed to the locker for our chewing-gum supplies, in order to serve out each man's chewing allotment, I found that evidently the crew had been unable to resist the temptation, for there was none left.

By this time Worrall had located the position of the leak, and Bonnett, always a handy man, was endeavouring to stop it with his only tablet of toilet soap. This seemed to do wonders, whereupon I decided that as a temporary measure soap was the thing. Clambering up on the top deck I saw a motor-launch coming alongside with the Senior Marine Officer and the Commissioner on board to greet us, whereupon I immediately hailed them with the request for soap.

They seemed a little surprised at this extraordinary greeting, but being good fellows they did not worry about such little details, and quickly grasped the urgency of the situation, for soon we were racing over the water to the landing-stage, where the hue and cry was set up among the natives, who dashed off to the village with cries of " Soap! "

Very soon bars of it were forthcoming, and quickly we returned to the Singapore, where Worrall forced solid blocks of this soap against the formers in the hull-bottom, temporarily preventing the inrush of water. By the time that we had the leak checked and well in hand, with the hull pumped moderately dry, it was quite dark.

I decided that it would be necessary to keep reliefs pumping the water out at intervals all night, and the Commissioner arranged for six natives to do this job by working in pairs in three reliefs. At the same time I knew it was essential for one of our own men to be on board also.

In view of the fact that Conway was responsible for his engines, and was single-handed in this matter, and that Worrall was going to attend to the refuelling, Bonnett volunteered for the job. He told me afterward that there was no chance of sleep all night in the Singapore hull, because each pair of natives would discuss between themselves the politics and general affairs of the day. One of them was quite convinced that these flying-boats were only a makeshift after the War, and the other agreed, seeing that all the affairs in Europe were upside-down!

The Commissioner and the Chief Marine Officer had journeyed all the way down from Port Harcourt to greet us, and that night, in the old, disused rest-house, they prepared for us, in the most wonderful style, temporary accommodation. Mrs Whitman, the Commissioner's wife, was chiefly responsible for this, and after all our trials of the day we finished up with a delightful dinner on the open veranda, illuminated by the light of the moon, with the stars as a ceiling.

Our flight from Bonny to Lagos was for the first part over a mass of creeks and lagoons that form the many mouths of the river Niger. In fact, the entire trip to Lagos was over a lagoon-country, where it would have been possible to land our flying-boat at any moment in ideal, sheltered waters.

From the air Lagos presented a wonderful spectacle, and the moment we circled over the native quarter of the city the streets became a black mass of rocking

LAGOS HARBOUR FROM ABOVE

TAKORADI HARBOUR AS WE FIRST SAW IT

humanity. I was told afterward that as we circled round, so the people ran up and down the streets in an endeavour to keep pace with us, and, as one chief remarked, a hundred thousand turned out to greet us and only one was killed. He, poor fellow, had stepped back into a passing motor-car.

. After a marvellous civic reception in the harbour, where the excitement with our mooring arrangements had been so intense that several natives fell out of the rowing-boat that was in attendance, Worrall and I set about to find a suitable crane to lift the Singapore so that we could try to trace the leak in our hull. The harbour authorities were exceedingly kind, and despite the heat of the day and the lateness of the hour they turned the dockyard workshops upside-down in order to find metal to make a spreader, and an odd cable to form our lifting-tackle.

The next morning an old steam crane lifted us some twenty feet from the water and swung us over the jetty, where we rapidly set to work on the task of repairing the leaks in our hull. We were not long in discovering the loose rivets, and making a repair, but decided that while we had got her out of the water we would give the hull another coat of paint.

My chief trouble during this operation was to keep the native dockyard-workers from swarming all over the machine. Every boy that we had employed in either cleaning or painting seemed to have about six mates who were 'legitimate' helpers, and at one time I discovered about a dozen boys in the hull who had artfully contrived to get on board under the plea that they were working on the machine, whereas in reality they had nothing to do with it.

It was almost pathetic to watch their keenness just to touch the Singapore while she hung some four or five feet from the ground, and I feared for the general safety as thousands started to swarm to this particular jetty. Luckily I hit upon the idea of walking round the Singapore in a circle, touching on the face with a wet white paint-brush everybody who came near me. This had the desired effect, for they scampered and fought right and left to escape getting the white dab. They took it good-naturedly and thought it a huge joke, and three or four who were standing with their backs to the jetty rocked with laughter so heartily at seeing one of their mates getting the white paint-brush full on his nose that they rolled backward, lost their balance, and tumbled some twenty feet into the water below. This caused greater merriment than ever. I was glad enough when the job was done, so that we were able to lower the Singapore into the water just before sunset and take her away up to her moorings in the middle of the harbour.

During our stay my wife and I were the guests of the Chief Secretary, Sir Frank Baddeley. On this particular night he had given a special dinner in our honour, and I disgraced myself by falling asleep over my soup. I think it was from sheer exhaustion from the day's work lifting our flying-boat, mending the leak, repainting her, and getting her back again.

To fall asleep over dinner is an old offence of mine, and my wife had saved me from disaster on more than one occasion by directing at me across the table severe looks that warned me of the danger, so that with superhuman efforts I fought sleep—always one of my most deadly enemies.

After a rest of three days we left Lagos and continued on our flight along the west coast. Leaving Nigeria behind, we passed the French colony of Dahomey, and came to the Gold Coast, a British possession which is sure to increase in importance in the near future.

We passed over Accra, its port and capital, being unable to land owing to the fact that there was no harbour there, the port being an open roadstead. So, after circling once, we continued to our destination, the great new harbour of Takoradi, just beyond Sekondi.

This great undertaking had been carried out by building two giant jetties out into the sea, forming an artificially closed harbour about a mile long and half a mile wide—a magnificent enterprise.

The port had only just been opened by the Right Hon. J. H. Thomas a few weeks before, and we were rather proud of the fact that ours was the first flying-boat to land in this great harbour, which it is believed will prove to be the foundation-stone of commercial prosperity in the Gold Coast Colony.

Chapter IX

THE IVORY COAST

OUR flight along the British colonies of the west coast was unique because it was the first time in history that a British aircraft had been seen in those parts. There was one story current regarding three natives in dispute, one very pro-British, and the other two not quite so certain about it. On the arrival of some foreign planes on the Gold Coast the latter two chided the pro-British native, asking him if Britain had any aircraft at all, and being nonplussed he replied, " Of course they have, but they don't go in for little things like these," and he pointed to the three scout machines that had arrived. " In England they only go in for big machines, ten times that size." He then went on to boast of the magnitude of British aircraft, and, as the Commissioner said, " let his imagination run," for he knew nothing about the subject. So when we arrived with what was at the time the largest all-metal flying-boat in the world our pro-British native swelled with pride. As the Commissioner remarked, " Thank heaven, you did not come in a light plane, or the poor fellow would have been in tears."

The heat in the harbour of Takoradi was terrific. Somehow it is sheltered from the cooling breeze, and it was like landing in an inferno. However, the reception that we received, and the hospitality extended to us afterward by the Commissioner of the Western Province, the Hon. H. E. C. Bartlett, soon made us forget that little discomfort. My wife tried bathing here, but the water was so warm and sticky that she had to bathe again to get it off.

We had no trouble in getting off inside the harbour, and before we were half-way across its smooth waters the Singapore was in the air, and finally cleared the breakwater by a hundred feet.

Our arrival at Abidjan, the next stopping-place on the Ivory Coast, was beset with petty disasters, but on reflection we found it to be a most humorous affair.

First we touched the bottom in picking up our moorings, because they had been placed at a spot where there was insufficient depth for our flying-boat. The moment this happened our new hosts were very disturbed, and instantly started a furious argument among themselves as to who was to blame for putting the moorings in that position. In the meantime we had made fast to these moorings, which instantly broke, causing a fresh outbreak of argument on the launches. In the meantime Mr Lewis, Elder Dempster's representative, appeared on the scene, and forthwith volunteered to go and get another anchor, while we, on the other hand, began to assemble our own, and get it out on deck ready to heave overboard.

There was not a breath of wind, or any current, so

there was no immediate hurry. I decided, in these circumstances, to wait for Mr Lewis's return. His anchor, we surmised, would be more substantial than our own, which, incidentally, stood about five feet high and weighed some 70 lb., having an enormous sharp-edged flux, with a wonderful holding power.

In a few minutes back came Mr Lewis's launch full steam, while that gentleman stood up in the bows waving in his hands what appeared to be in the distance the end of a boat-hook. He was soon on top of us, shouting, " Here you are ! " While Worrall and I were struggling to lift our seventy-pound anchor Mr Lewis appeared on the scene and presented us very proudly with a miniature grappling-iron that might have belonged at one time to a first-class model yacht on a London pond ! The terrible part about it was that he was in dead earnest, and at the time I almost felt that I wanted to hide our own anchor.

When I had asked Mr Lewis to bring us a big anchor I had added that we should be grateful for a drink and possibly a sandwich, but all things in life are purely a matter of perspective, for if Mr Lewis's anchor was small his lunch was big. We soon discovered that the drink consisted of crates of beers, sodas, and spirits, and the sandwich consisted of baskets full of them, and containing enough food to feed a battalion, let alone light refreshment for the crew of the Singapore.

We had made fast, and were having a little lunch prior to refuelling immediately, when I noticed an old tub of a steamboat slowly drifting down on to us.

There was nobody on board whatsoever. We were all alarmed at this terrible phenomenon, which, it seemed, was going to wreck us, when I espied under our tail three or four natives in a rowing-boat quietly inspecting our machine. It was Lewis who realized what had happened and started to hurl curses upon the party in the boat, for it appeared that they were the crew of the steamer, who had just abandoned their ship in midstream in order to do a little bit of sightseeing.

I do not know whether it was the fear of going to Hades, or of being shot or flayed alive, but, anyhow, Lewis's vocabulary in French and mine in English (which they did not understand) made those natives realize that if they wanted to live they would have to prevent the old tub from touching the flying-boat, and they somehow got on board in the nick of time. The terrible part of it was that she was drifting broadside, so that when they went full steam ahead they almost rammed us, and narrowly missed taking off our nose. After this little episode I was at pains to explain to all present that the real excitement of a flying-boat cruise was not in the air, but after one had landed on the water.

Later that afternoon we attended a magnificent *Vin d'Honneur* given by the Governor, and on this, as on many other occasions, I deeply regretted my inability to express my feelings of gratitude to our French hosts in their own language. French has never been my strong point. However, my wife, who was the linguist of the party, came once again to our rescue.

Our next journey was to be, possibly, the longest non-stop jump of the whole trip, a voyage of over 700 miles, with the likelihood of a head wind. There was no place, with the exception of Monrovia, on that lonely stretch that covered the whole length of the Liberian coastline where we could land amid habitation and life. We had a magnificent take-off from the lagoon at Abidjan, and had soon settled down to our long flight. The wind was negligible, the day was fine, and all was going well, when I happened to go below, and discovered that the gauges were registering very high temperatures, which caused anxiety among the crew. We had done about eighty miles when I could feel undue vibrations, and, leaving Worrall at the controls, I again went below. After a little consultation with Conway I came to the conclusion that something serious had gone wrong.

I studied the charts and discovered that the configuration of the coastline that we had been following for over 1000 miles was going to make a sudden change, for instead of a perpetual string of lagoons of calm waters running parallel to the coastline, on which we could alight at any moment, we were coming to a hilly and rocky coastline, harbourless, with giant breakers, and lacking any refuge whatsoever for a flying-boat. We were just about to leave the last lagoon behind us when the sound of my engines made me decide to land immediately.

As soon as we came to rest on a vast open water, for the lagoon was several miles in length and about a mile in width, Conway got to work and made an

WE LANDED ON WHAT APPEARED TO BE AN UNINHABITED
TROPICAL LAGOON

HOW THE ROOTS OF THE TREES WERE LEFT HIGH AND DRY

examination, and being unable to trace any trouble we decided to try again. We took off, and after I had circled once I landed again, being dissatisfied with the running of the engine. Another long inspection followed, as a result of which it was politely suggested that my fears were perhaps unfounded, so we took off once more.

This time I had hardly got fifty feet in the air when I came to the conclusion that it was not good enough, and just landed straight ahead. On the third investigation Conway allocated the difficulty. At first we did not realize how serious was the situation, but gradually it dawned upon us that, as we should have to send to England for our new part, this might mean a delay of weeks. The crew became despondent, for there we were in the middle of a lagoon, whose shores were a mass of dense jungle, and without a living soul in sight. We knew that we were about fifty miles from any form of communication, and some of the crew, I think, were afraid that we were on an uninhabited part of the coast, and that we should be unable to obtain food or proper water.

We had three or four days' emergency rations on board, but pure water was the difficulty. We always looked upon the water in the radiators of our engines as a supply to be tapped in the last resource, but that particular water would be mightily unpleasant to drink.

Presently we discovered that the place was not uninhabited, for a solitary native appeared, standing erect, with punting-pole in hand, balancing himself in the smallest canoe I have ever seen. The canoe, which was about twenty feet long, consisted of a tree-trunk that had been hollowed out, but it was not more than fifteen inches wide and about six inches deep. I beckoned to him to come alongside, but at first he was very reluctant to do so, and as he could not speak or understand English or French we were unable to ask any questions.

All I wanted to do was to get in touch with civilization, and suddenly I remembered that a few miles back we had passed a steamer that appeared to be anchored a long way out to sea. We had presumed that she was taking on an odd cargo of mahogany from some native village, and I clung on to the hope that this might be our link with communication.

We discovered that the native knew what "big boat" meant, for he pointed down the lagoon in the direction from which we had come, and as the lagoon ran parallel to the sea we surmised that he knew all about the ship in question. We had already dropped anchor, and as it was no use wasting time I decided that I must get in touch with this ship if possible. After many signs and much use of the expression "big boat" I persuaded the native to take me in his canoe in that direction.

I think my wife was a little nervous of the idea of my going off with this all but naked native in a perilous-looking craft which was already half full of water, and she insisted upon my taking my revolver with me. It was quite a balancing feat to sit in this more or less hollowed-out plank without upsetting it,

CONWAY INSPECTING ONE OF THE CONDORS (700 H.P.)

and soon we were shooting across the lagoon at an alarming speed, considering that my native friend had to balance himself while he stood up and manipulated a long, unwieldy punt-pole.

In half an hour we came to the end of the lagoon, and then shot through a little cutting that brought us into another lagoon that was separated from the sea only by a sandbank. A few moments later the ship was on the horizon. Soon I could hear the thundering of the mighty surf, and presently the chanting of natives working in unison as they rolled five-ton mahogany logs down the sandbank into the breakers, to be dragged out on a line to the ship. My native put me ashore on to the sandbank, and, jumping out, I ran up the slope. On reaching the summit I came upon some hundreds of natives who were at work log-rolling, and from the beach to the ship, which was over a quarter of a mile out, there was one long string of logs tied together in single file. However, the thing that cheered me most of all was the fact that there was a white man among them, and, rushing forward, I discovered that our new friend was French and spoke no English. My French is pretty bad, but immediately we got on famously, and in a few moments I discovered that we should be able to send a wireless message from the ship, and that night we should be able to sleep in his hut on the hill on the far side of the lagoon. Furthermore, that we should " manger à huit heures."

I decided to send a wireless message off to Freetown immediately, and going into a small hut I wrote it out in plain block letters, and after putting it into an envelope handed it to a native. Then for the first time in my life I saw a surf-boat launched, and waiting its chance to get through the west-coast breakers. It was a thrilling spectacle, and it seemed as though the boat would never rise in time to ride the crest of those twenty-foot waves.

In the meantime I learned from M. Pierre that we could make a journey through the canals in a dugout canoe to Lahou, some fifty miles away, and that by starting at dawn we might reach our destination before dark. With this information I started back in my canoe to the flying-boat because I knew that they would all be anxious if I was absent too long; in fact, when I did eventually arrive on the scene it was a great relief to them.

Our original fears that the lagoon might be uninhabited were now dispelled completely, for there were literally hundreds of natives in their dugout canoes swarming all round the Singapore.

Quickly I explained the situation to the rest of the crew. We were in a lagoon at the back of a native village called Fresco. It was not a port, but one of those places on the coast where ships called occasionally to pick up cargoes of mahogany. Therefore we were lucky in finding a ship to send off our message, so that Freetown would not be worried owing to our failure to appear on time. As the coastline of Liberia was considered treacherous from every point of view any delay would naturally cause anxiety.

It was evident that in order to get our supplies from

KRU BOYS SPORTING IN THE SURF

SURF-BOAT COMING IN ON TOP OF A BREAKER

England I should have to be in direct cable communication, and, as far as I could see, the nearest place from which I could do this was Lahou. Furthermore, although M. Pierre might have food enough for himself, six more of us would soon exhaust his supplies.

For the time being we left Conway and Bonnett in charge on the flying-boat, while my wife, Worrall, and I, having found a bigger canoe, set off for Fresco again.

Later that evening M. Pierre gave us a very good dinner in his grass hut. We learned that on the following day he was going up-country to inspect some timber, and he invited us to use his hut during our stay.

After a little consultation we made the following plans. It was essential that some one should go to the nearest point of civilization, not only to get into communication, but also to receive supplies when they arrived from England, and bring them back to Fresco, which was not a port of call. It was agreed that Worrall should stay behind with Bonnett and Conway, and should be responsible for the well-being of the Singapore.

Arrangements were made that my wife and I should set out the following morning to go by canoe to Lahou, from whence we would send food and provisions back to the boys on the Singapore.

We also made arrangements with the local chief for about twenty canoes to tow our flying-boat round from the inner lagoon to Fresco in the main lagoon by the sea. This operation was going to be done at dawn on the following morning before the wind came up, and after these decisions we said good-bye to Worrall, who went back to the Singapore in order to be on the job early in the morning.

The night was hot and sultry, and the mosquitoes drove us mad in the grass hut. In addition it was so terribly close there that my wife and I, enveloped in mosquito nets, sat out in the moonlight trying to sleep, but in reality praying for the dawn.

The sun rose and the mosquitoes went, but the sunshine was of short duration, for by the time we had got our baggage on board the big canoe huge black clouds were coming up over the horizon.

Our canoe was forty feet long, about a yard wide, and of the same depth, a solid log of timber having been hollowed out to make the craft.

M. Pierre helped us wonderfully. He arranged with the local chief firstly to supply us with the canoe, and secondly to find a crew for it. Our crew consisted of about sixteen stalwart natives, who were armed with a paddle and a punt-pole apiece. There were about eight in the bows and eight in the stern, and my wife and I sat in the centre.

We were instructed by the chief on no account to let the natives eat, sleep, or stop paddling, and finally we pushed off as a tornado broke, to the accompaniment of thunder and drenching rain.

For hours and hours we wended our way through narrow canals bordered by dense jungle forests, through which the sun could hardly penetrate, and

OUR HOME FOR A MONTH ON THE WEST COAST

OUR CANARY KEPT US CHEERY UNDER THE TRYING
CIRCUMSTANCES

then came a period when the one and only waterway that constituted the connexion with Lahou passed through a swampy area. The storm had been of short duration, and now the sun was burning fiercely, but fortunately, before starting, we had bought at a native store at Fresco two umbrellas at one and sixpence each. This was a line they ran specially for the local market. These umbrellas made fine sunshades as hour after hour we pushed our way through the rushes that fringed a canal which was often no more than a bare yard in width.

Eventually we came to another native village, called Big Town, situated at the western end of an enormous lagoon. We had been in the canoe over six hours, and all that time we had gazed on the shiny ebony backs of our native crew, as they paddled in rhythm hour after hour to the chant of one or another of them, while in the stern the head boy had steered us, among shallows and under overhanging trees, through narrow waterways in the swamps, and at length had brought us safely into the big lagoon.

Here, as a result of a further wireless message from the ship, a petrolette was waiting to take us on to Lahou. On the Ivory Coast any kind of craft into which they may happen to put a motor is termed a petrolette. Ours, as it happened, was a good one, and soon we were speeding over the lagoon at four or five knots, having said good-bye to our native crew who had laboriously brought us through the shallows at some two knots an hour.

This particular lagoon seemed unending, and hour

after hour we continued, only managing to get into Lahou by dark.

On arrival at Lahou we were received with open arms at the establishment of Wodins, an old British trading-house whose history dates back for centuries. Here we found Mr Ogilvie in charge, and with him his wife just out from England. However, we could not stay very long with the Ogilvies as the *Administrateur*, M. Casanova, was waiting to receive us.

That night a terrible thunderstorm washed away the telegraph line, and still we were unable to get in touch by cable with England. In the middle of these affairs Mr Lewis from Abidjan burst in upon us. Our first wireless message had been picked up from Fresco at Bassam, his headquarters, and forthwith he had set out by petrolette, and had been journeying day and night until he met us by surprise at Lahou.

Immediately he decided that we should have to go back with him to Bassam, as that was the port, and furthermore we should then be in direct communication with Europe. Foodstuffs and supplies were sent off to Worrall, while we journeyed eastward toward Bassam, which was quite 120 miles from where our boat was resting at Fresco.

Mental depression set in badly at Bassam, for despite the wonderful hospitality of our friend Lewis there was the overpowering thought of having to wait a month for our stuff to arrive from England, and of how our delay would imperil the success of our flight round Africa.

Once the excitement of the cabling was over, we

THE EVER-NARROWING CANAL

PUTTING THEIR BACKS INTO IT

faced the situation. My wife and I set to work to write this book, and then at four o'clock in the afternoon we took our exercise by playing tennis on a very old and disused concrete court. At first we played alone, but soon the French community began to join in; in fact, we started a tennis craze, and one lady who lived in Abidjan seemed very keen that we should go over to that town in order to get the movement working there.

At last the day came for the boat to arrive from England. Soon we were all packed up ready to start on the 120-mile journey through the lagoons back to our machine at Fresco.

Evidently those at home thought we were all fagged out, so they sent us two more crew to help us, and the moment they came ashore they were so keen to lend a hand that I dispatched them off immediately on the first canoe for Fresco.

Lewis insisted upon coming with us all the way to Fresco to see us off. Having had a terrible rush to clear up everything at Bassam, we had made a late start, and were six hours away from Lahou when darkness overtook us. We had a perilous journey down one canal in the dark, for we met a great big motor-barge coming in the opposite direction, and were unable to attract their attention. There was room to pass only by both getting well to their respective sides of the canal. We had already come to a standstill right up against the rushes, and without altering its course this 100-ton barge was bearing down upon us, when at the last moment the sleepy native in

215

charge saw us. It was too late to clear us completely, with the result that we were jammed in between the bank and the barge, and had it not been for the fact that the rushes gave way we should have been crushed and smashed to pieces.

Our adventures for the night were not yet over. It was midnight, and we were in sight of Lahou and nearing a tricky part of the journey. It was where the lagoon ran into the open sea, although there was no passage through to the open, owing to a big sandbar which was just awash at low tide. By keeping well away from the sea this bar could easily be avoided. Suddenly I looked over the side of the boat and noticed that the water was all white, and the next instant there was a crunching sound: we had run hard and fast on to the bar. Instantly the boat rose up on to her keel, and all but capsized, while our provisions and crates of minerals and crockery were upset and smashed to pieces.

All of us clung on expecting every moment to be flung into the sea, but fortunately the tide was coming in, and with our engines turned off we soon drifted off the bar and cautiously retraced our course.

Presently we met a local native, who, after a few words, came on board and piloted us in the darkness, through the safe channels, into the back of Lahou.

An early start was made for the last stage of our journey to Fresco, but on arriving at the native village of Big Town, where we were to change over to the canoe in order to get through the shallows and swamps, we discovered that no paddles were available.

Eventually I persuaded the local chief that he was the greatest chief in Africa, and that I wanted to be able to tell the people in England how that when there were no paddles he, with a wave of his hand, had commanded paddles, and immediately they had come from nowhere.

Of course all this had to be translated, and possibly the interpreter had embroidered my story a little. At any rate, the effect was marvellous. The old boy went round his village, and everybody had to contribute a paddle, so that eventually we were all ready to start once more.

However, the paddle question had caused serious delay, and again darkness overtook us before we reached our destination, but this time we were in for a most terrifying experience.

The one and only canal into Fresco winds its way for many miles through a deep, tropical, forest jungle. It was a dark night, unrelieved by the light of moon or stars. Even in the open the blackness rendered visibility *nil*, but when we entered the forest the inky blackness was such that it was literally impossible to see your hand an inch in front of your eyes.

All day long we had had to urge our crew in order to keep them moving at all, so therefore now we were expecting them to stop completely, but, instead, the darker it got the faster they paddled. One native, who had done nothing all day, sat right in the bows and looked ahead into the blackness. He yelled orders to the man in the stern, who steered accordingly, and then when we came to the thick forest

217

everything was blotted out, and we could hear nothing but the shouts of the man with the cat's eyes, and panting breath as our crew paddled faster and faster, plunging on under tree-branches into the darkness. I was really afraid, for if we should be overturned, apart from the fact that these waters were alive with crocodiles, there was no foothold on the bank or shore because we were in the swamps, and furthermore we could see nothing.

We were only too thankful when at last we came out of the forest and to our joy could see a starlit sky.

Two days later we were ready to continue our flight of survey, our only difficulty being the take-off.

Since we had arrived the natives had let the water out of the lagoon into the sea, as the waters were flooding the village. The result was that the big lagoon in which we had originally landed was now too shallow for us to go into, so we had to take off out of the small lagoon which ran parallel to the sea.

From a canoe I sounded the lagoon from end to end, with a piece of iron tied on the end of a line. There was just over half a mile run, and at points it meant steering within two or three feet of shallows that would have wrecked us if we had touched.

On the first attempt to take off we could not get unstuck. On the second attempt we hung on to the last moment, and when there were only a few more yards to go before we reached the extreme end of the lagoon I was just about to shut off when she lifted a little, and instead I managed at the very last moment to pull her right off.

ARRIVING OVER FREETOWN

THE HULL OF THE SINGAPORE BEING CLEARED BETWEEN
THE TIDES AT FREETOWN

Chapter X

FROM FREETOWN TO CAPE BOJADOR

THE joy at being in the air again after our month's delay was wonderful. Everything was going well, and instead of meeting a head wind as we had anticipated, as we flew north-west the wind was on our port beam. We were making over 100 miles an hour, and after a 600-mile journey we sighted Freetown, the capital of Sierra Leone.

The journey had been a joyous one, because we had passed over a long stretch of bad coastline without difficulty, and on arriving at Freetown we came down on to wide, spacious, calm waters.

No sooner had we made fast than I discovered that my crew down below were busy getting out bilge-pumps, for we had sprung a leak again. I guessed that perhaps corrosion had set in as a result of the Singapore being at anchor in one position for a whole month in the foul waters of Fresco Lagoon.

I explained these details to the harbour-master, who instantly suggested that I should send some natives to dive underneath the hull to inspect the Singapore. In a few minutes we had stories of great shell-fish and jagged edges all over the hull, which accounted for our bad get-off from the Fresco Lagoon, for such friction on the hull's surface would destroy immediately its power to slip through the water.

We were all dead-beat and wanted to get ashore. We hankered after the pleasures of civilization—such as a good bath, and all those things that we had missed on the wilds of the Ivory Coast. But I could see that we should have to wait for these things, and so Worrall and I made a tour of the docks in search of a crane that could lift the Singapore. We could find nothing with a reach big enough to do the job. At this juncture Mr Archer, the manager of the Elder Dempster Company, suggested that we should go round to their engineers' yard, which was up a creek, and here we discovered what is known as a ' grid ' for cleaning barge-hulls.

The grid was formed of huge timbers lying length-ways across one another on the mud. The whole structure was built on a firm base, so that it was possible at high tide to bring our flying-boat over the top of this grid. When the sea went out the keel of the Singapore would sink on to the timbers, and by chocking up the hull all round she would be left high and dry for a gang of native workmen to get underneath and clean the hull.

It required absolute high tide to get the Singapore over the grid, and on making inquiries we found that our opportunity to do this job would be at five o'clock the following morning. Having put guards and reliefs on board we all went ashore, and only too gladly accepted the warm hospitality extended to us. The Colonel of the regiment took charge of all my crew, while my wife and I were the guests of his Excellency the Governor, Sir Joseph Byrne. That night we went to bed terribly tired, so that when on the following morning we had to get up at four o'clock in order to catch the tide it was literally cruelty to human beings. Nevertheless we were all up on time, but somehow the towing arrangements went wrong, and by the time we got round to the creek the tide was already running out, and we were too late to get over the grid, with the result that we had to go back to our moorings and wait until the next day.

On the following morning we took no chances, and were up long before dawn. As high tide was a bit later we had ample time to get our flying-boat right over the grid, and then we waited until the water lowered her on to it. It was a ticklish job, because certain sections of the keel had to fit within a few inches on to particular logs, and in addition wooden chocks had to be in readiness to block up the chines of our hull in a most precise manner as the tide went out.

Robert Price, Elder Dempster's chief engineer, came to our aid in this affair, and got into a bathing costume for the job. Worrall and Bonnett followed his example, and as the hull sank she was very carefully guided on to the correct beams, and wedges were placed at the right spots on her hull to prevent undue strain from being placed on portions of her structure that were not meant to take the weight.

As the level of the water sank it exposed a mass of barnacles and shell-fish, and around our stern was a skirt of seaweed; in fact, with so foul a bottom it was a wonder that we ever got off the water at Fresco.

Busily an army of some fifty natives worked at scraping our hull clean, while we endeavoured to make good all the corroded rivets; in fact, we worked solidly for nine hours without intermission, until the tide began to overtake us. The work on the hull continued until there was only about a foot of water between the keel and the heads of the men that were working.

We were seriously delayed in our work by a terrible tornado which sprang up and blew violently down the creek, broadside on to the Singapore, and I feared that every moment she would be slewed off the temporary chocks that were supporting her and crash as a consequence. I was glad enough when the tide came in so that we were able to refloat her and take her out to her moorings in safety.

The time at Freetown had been three days' hard labour, more especially as the two additional men who had been sent out to help us had to go into hospital immediately on their arrival. On their long canoe journey from Bassam to Fresco through the lagoons they had unwisely exposed their knees to the sun, with the result that they had each got a bad attack of sunstroke. Their legs became burnt, and when they reached Freetown their knees were, without exaggeration, quite three times their natural size.

Thus our extra helpers had been more or less invalids and an additional anxiety to the expedition ever since they joined us. I believe we all overworked at Freetown, and I think that I personally suffered much from the continual drenchings that I got in the tornado downpours. As the job could not be left our shirts and trousers dried on us in the heat of the sun, only to be drenched again in the next downpour.

Our take-off from Freetown was a bad one, for it so happened that there was a low, rolling swell which gave our flying-boat a tendency to porpoise. However, on the second attempt we got well away, and we were soon heading the nose of our machine northward for Bathurst, our next stop.

. On this journey we noticed a marked difference in the nature of the country. Tropical vegetation gave way to bush, and by the time we reached the old British colony of Gambia we were flying over scrub.

The colony of Gambia, broadly speaking, is that territory which borders the banks of the river Gambia for about 250 miles from its mouth; and Bathurst, which is the colony's port and capital, is some twenty miles from the Atlantic. Soon after landing we went ashore, and here we discovered that a great reception was awaiting us. The Governor and Commander-in-Chief, Sir John Middleton, K.B.E., C.M.G., had been very ill and was staying at a country residence on the coast during his convalescence, but on sighting our flying-boat he had journeyed into Bathurst specially to greet us, and, after staying a brief hour, under doctor's orders had to return again to his villa on the coast.

After the reception I wanted to send a message back to my men on the Singapore, so, having gone down to the quayside to superintend the dispatch of the fuel, I handed the note to a native marine who was in charge of the motor-boat, and who set out at full speed to carry out the mission. His approach to the flying-boat was to the stern, and Bonnett told me afterward that he was aroused from his particular task at the time by shouts from the native in charge of the launch. On looking up he saw this motor-boat about a hundred yards off coming full steam for the starboard beam. Our energetic marine was standing up in the boat waving a letter and shouting, " I'm coming, Boss, I'm coming." For a few seconds this was appreciated by all on board, but when the launch was within about sixty yards and there was no sign of any abatement in speed my men naturally became alarmed, and all started to shout to the oncoming launch to go steady. The only effect of their cries was to provoke renewed shouts from our native marine, who joyously cried with a broad grin on his face, " I'm coming, Boss, I'm coming." The native in his excitement suddenly realized that he ought to put his engine into reverse, and gave a quick order to his companion in the boat to do so, but the gears did not take effect. In that moment his expression changed to one of fright, but even so he continued to shout, " I'm coming, Boss, I'm coming. I'm coming, Boss, I'm coming." *Crash!* and the bows of the pinnace cut their way deep into the trailing edge of our lower plane, until with a mighty thud they came up against our lower main spar. When the impact flung the enthusiastic native headlong on to the Singapore his first words were, " A letter, Boss."

I do not believe that my men were very complimentary at the time. At any rate, they were far too busy examining the broken plane and trying to ascertain whether any damage was done to our main spar to care about him or his message.

Somehow the spar had withstood the shock unharmed, but it took two of my men a whole day to repair the damaged plane. It was an awkward job in a skiff on choppy water, as they worked with their arms above their heads under the lower plane. Every now and again you would hear them mutter, " I'm coming, Boss, I'm coming."

We received a cable from the British Consul at Dakar, who was very keen that we should pay that port a visit, but the lack of any satisfactory information of the mooring possibilities, combined with the fact that we were long overdue, made me reply reluctantly that we should be unable to accept the kind proffered hospitality of our French friends.

As we flew northward we cut across the land over the peninsula, leaving Dakar and Cape Verde well away on our west, and very soon the country had become absolute desert. We passed Saint-Louis, the great port at the mouth of the river Senegal. Here the heat was so great that, despite the fact that we were some four or five thousand feet up, the temperature in our hull was nearly 90 degrees.

We were making for Cape Blanco, where we were going to land in the lee of that long sand-spit peninsula at a little place called Port Etienne. It is a French port on the frontier of the Spanish Rio de Oro, and here our supplies had been laid down months before.

After passing Saint-Louis, for nearly 400 miles, until we sighted Port Etienne, we could see no sign of life whatsoever—not a track, not a blade of vegetation, but just a long coastline ahead of us, with the endless golden sand-stretches of the limitless Sahara away to the east, while away on the west were the vivid blue waters of the Atlantic, the only sign of movement being the white-crested foam of the surf on the beach.

Although the air appeared to be comparatively calm at 5000 feet, as we came down to a lower altitude on approaching Port Etienne we discovered that there was a violent wind blowing off the land from the north-east. The peninsula at Cape Blanco afforded perfect shelter from the Atlantic rollers, but, even so, we landed on a stiff chop head into a 40-mile-an-hour breeze.

I don't think that we realized the force of this wind at the time, for, having approached our moorings and picked them up with a light line, the engines were shut off and I went forward to give Worrall some assistance. At that moment the flying-boat started to drift astern with such force that we were unable to pull up on to the buoy. I looked behind and discovered that our direct drift would bring us on to some rocky cliffs on the peninsula. Just at that moment, somehow, the rope slipped off the bollard, and Worrall, in trying to make it fast once more, got the rope caught round his body, and in a second the whole weight of our craft in the 40-mile-an-hour wind was being taken by the rope, which was almost cutting Worrall's body in two. Worrall, in his endeavour to get free, twisted round, but a further gust of wind brought the bar ends of one of our bollards into his back.

My shouts for assistance caused the entire crew, including my wife, to come on deck, and then we found the difficulty of getting a hand on the rope ahead of him in order to release him. This, however, was finally accomplished by releasing the whole line, whereupon Worrall slipped out of the loop.

Then came the difficulty of hauling our machine up on to the buoy once more in order that we might get our main mooring-lines attached. Owing to the terrible strain on the one line that was holding us, we could see that the strands at one point were giving, and that unless we were quick, and hauled the line in beyond the bad part, it might break at any moment, and we should drift on to the rocks and be wrecked.

There were seven of us on the line, and it seemed as though we should never heave her in. The pull on the machine was so great that we dared not let the rope slack for a second in order to take a twist over the bollard, let alone heave in and capture the breaking strands.

I urged and implored the men to pull for all they were worth, and my wife's hands were being torn in her desperate endeavours to help to save the ship.

Those on the shore could not come near us because of the gale, and at one time it seemed as though we should have to give in from sheer exhaustion. Gradually, however, we began to gain inch by inch, until finally we had regained the snapping portion of the line. Soon we were hard up on to the buoy, and then, leaning over, we hooked another stout line through the ring of the buoy, and at last made fast.

By this time everybody on board was almost exhausted, but the worst was not yet over. In fact, our troubles had hardly begun, for the next difficulty that presented itself was that of getting ashore.

Port Etienne was an outpost trading-camp and possessed no motor-boat, and to row against wind and current was an impossibility. I do not think that our friends ashore knew anything whatsoever about flying-boats, for they took it for granted that they could come alongside in a sailing-boat. Of the danger of getting their masts entangled in our propellers, and of the possibility of getting their boat jammed under our lower planes on the tossing waves, or of the fact that our metal hull was only a thirty-second of an inch thick and not meant to stand the buffetings of the steel bows of a sailing-boat, they seemed to be totally unaware.

In such a gale it was impossible for me to explain in French that my lower plane and wing-tip floats would not be able to withstand the heavy buffeting of their sailing-boat. I screamed and implored them to go away until I could think out some other method of doing the job, but aboard this sailing-boat there appeared to be one who spoke good English. His chief anxiety seemed to be that the whole reception would be spoilt, and when I tried to explain that we should certainly be wrecked, with a shrug of his shoulders he seemed to regret the inevitable. We were almost in tears by this time, and begged them to go away from us, while we frantically endeavoured to free their mast, which by this time had got jammed up behind our propeller. Having visions of the sudden and imminent termination of our whole expedition, I shouted the question, "When is this gale going to stop blowing?" and the reply came back, "Never. It blows like this day and night, year in and year out from the same direction."

Finally, in desperation, I jumped on board the sailing-boat and managed to push it clear of our wings, and soon we were drifting astern, clear of the flying-boat, which was safe from attack for the time being.

I then had time to think, and with the assistance of my French friend who spoke English we explained to some native boatmen that the only way to get on board the flying-boat in such circumstances would be to anchor off some distance up wind, and then from this anchorage lower down a small boat on to the flying-boat.

Very soon a firm anchorage and a buoy were put down about 200 yards ahead of the Singapore, and supplies of fuel were taken out in a sailing-boat to this point and so lowered down wind gently in a small boat on to our craft. The job was long and laborious, but we were greatly helped by the skill of the native boatmen, who soon began to handle their task with dexterity.

229

Long before we were finished my wife went ashore to pay our respects and apologies for our somewhat erratic arrival to the Commandant. Port Etienne is situated on a spit of sand jutting out into the Atlantic from the edge of a great desert. Westward the Atlantic stretches over 3000 miles to the West Indies. Eastward the Sahara Desert stretches over 3000 miles to the river Nile. Southward the nearest port or town of any description is 400 miles away, and northward the nearest port along the coastline is 1000 miles distant. The settlement itself consists of a small jetty, a warehouse in connexion with the fishing industry, and a small store, and two miles away is an aerodrome with a hangar on it—one of the refuelling stations for the French Air Service to Dakar. There is also a fort away on a mound, and this is where we were destined to stay the night.

My wife asked the owner of the store why his place was surrounded with barbed-wire entanglements, whereupon he replied that they were expecting a raid of a thousand Arab brigands at any moment, and proceeded to point out on the map the place where they had been sighted by the passing air mail on the previous night. It was estimated, he said, that they would arrive at the ordinary rate of march on the following day. When I was informed of the news I was determined that we should leave as soon as possible.

That night we all slept in the fort, and to our surprise we discovered that the Commandant had his wife and two children with him in this lonely spot.

Just before we sat down to dinner we could hear the clanking of door-bolts and the barring of gates as this little fort was closed for the night, and from the turret tower I could see the lights on our Singapore as she rode at anchor.

In view of the fact that the brigands were mounted men approaching us over land, every one thought that the member of our crew who was destined for guard that particular night was really in the safest place.

We could not but marvel at the hospitality and warm welcome of our host, who, despite the fact that we were really " out in the blue," gave us a dinner that would have done credit to a first-class restaurant.

Owing to the fact that, with the exception of a solitary Spanish fort on the estuary of the Rio de Oro, some 200 miles north, called Villa de Cisneros, there was not another port for over 1000 miles where we could land, we decided to fly out into the Atlantic to the Canary Islands.

I was a little worried about our next day's trip owing to the terrible head wind which we should have to encounter, it being the famous north-east trade. However, I was going to apply the same tactics as I had when we went down the Nile—namely, to try flying high to see if we could get above it.

When we started out in the morning the gale was still blowing so hard that it was impossible for us to turn down wind, and, in fact, inadvisable to attempt it. Thus, in order to get into a satisfactory position to take off inshore, head into wind, we had to drift backward, and then cut diagonally across the waves to our starboard. Then we drifted back again, continually repeating this manœuvre, making a kind of saw-edged zigzag, until we had drifted far enough into the middle of the bay for us to get sufficient run to get off the water before we reached the beach again.

As soon as we were off the water I started to take readings of our forward speed. During the first hour we climbed and flew at 1000 feet, and I discovered that we had covered just on sixty miles. We then climbed to 2000 feet, and having flown at that altitude for a quarter of an hour I measured up the distance covered on the chart and found that we were flying at the rate of 66 miles per hour.

We went up in steps in this fashion for 6000 feet, at which altitude we were making nearly 90 miles an hour, which proved that at that height there was no wind at all. I have been told since by meteorological experts that had we gone to 12,000 feet we might have met an anti-trade wind, blowing the reverse way to the ground wind—namely, from the south. Anyhow, it is comforting to think that aviators will have a method of getting over the trade-wind problem.

As we passed over Villa de Cisneros we looked down at the solitary fort, in which we were given to understand five men lived. It was the one place that was occupied in the whole of the Rio de Oro, a country with 700 miles of coastline, and running about 200 miles inland.

Chapter XI

THE CANARIES

AT Cape Bojador we left the coast and headed
our machine north-west for the island of Gran
Canaria, where we were going to land at La Luz, the
harbour for Las Palmas. At the very moment that
we left the coastline and started on our new course,
from our altitude of 8000 feet we could clearly see a
mountain on the horizon ahead of us. It was Pico
del Pozos, the peak of the Gran Canaria, which was
about 130 miles away. The visibility is so remarkable
in this part of the world that it is a common thing to
see with the naked eye for 150 miles. We passed
above the clouds for almost the entire journey, with
our objective in sight the whole time, until at last we
were circling over La Luz.

This harbour is on the east of the island, just north
of Las Palmas, and is formed by a jetty running down
due south from a promontory, so that it is open to
the south. The more or less permanent trade wind
in this part of the world comes from the north,
and so we landed into the mouth of the harbour.
Nothing could have been more nearly ideal as far
as landing was concerned, and an enthusiastic party
came out to meet us in a motor-launch, commanded
by the harbour-master, who proceeded to lead the

way up the mouth of the harbour, between banks of shipping.

Evidently they must have imagined that our flying-boat, with its 100-foot span, was as easy to manœuvre on the water as the harbour-master's little motor-boat. Very soon I found myself round a corner in the inner docks with the masts of ships in all directions, with only about 100 yards in which to lose way and come to a standstill in the dead-calm water, before hitting the dead end of the quayside steps.

Unfortunately, flying-boats cannot go astern, and our kind friends had not allowed for this. It was lucky that we had our drogues out, and these, acting as a brake, brought us to a standstill just in time.

Then, as a flying-boat is always a weathercock, we naturally started to head into breeze, and would have drifted back on to a big liner, and probably damaged our tail, but my crew by this stage of our expedition were fully aware of all the dangers, so that when I shouted to them to get ready to drop anchor I discovered that my wife had already been busy down below, and was struggling with this seventy-pound weight, handing it up to Worrall so that it would be ready to drop instantly if necessary.

The result was that simultaneously with my shout for the anchor there was a splash as it dropped over the nose, and as I looked down between my legs, through the medley of bars that made up the floor of the cockpit, I could see my wife easing the line as it paid its way out. When I saw our tail getting perilously near the liner I shouted to Worrall, " Make

THE CARPET OF CLOUDS BENEATH US

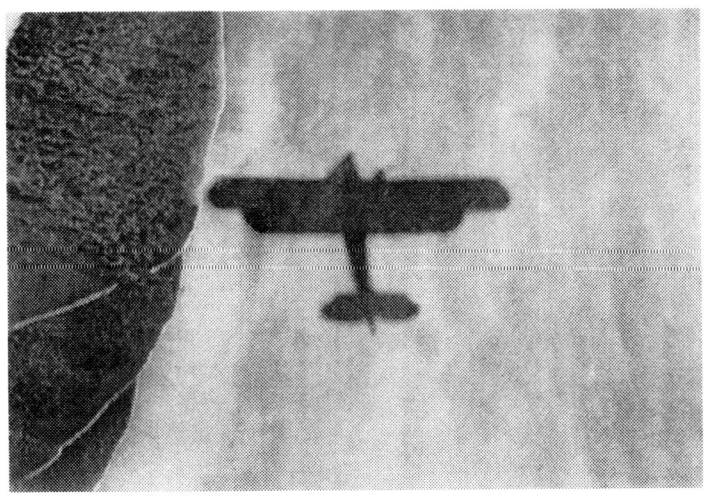

OUR ONLY COMPANY A SHADOW

fast," and it was a great relief to us all when the line tightened, the anchor held, and we came to a standstill.

When the mooring arrangements were finally completed, despite the fact that we were in civilization once more, surrounded with smiling faces and kind folk who wanted to greet us, curiously enough my first remark to Worrall was, "How are we going to get off from this place?"

To take off inside the harbour in the same direction that we had landed into it would be impossible. Firstly, we could not go out into the open sea to get a good run, owing to the permanent Atlantic rollers, therefore we could only go as far as the mouth of the harbour. Assuming that we were able to get off the water in this short run into the north wind, we should find ourselves with ships' masts towering on all sides. It would be impossible to climb over these in so short a space, and even if the ships were not there there were the tall houses of the town to get over and a high mountain straight ahead, to say nothing of the difficulties of negotiating all the down-currents.

I pointed these things out to Worrall, who replied, "Never mind, we shall have to take our time, and instead of taking off into a north wind go right into the north end of the harbour, and take off down the whole length of it toward its mouth into a south wind. Then, we shall have nothing to clear, just the open sea before us."

That sounded all very simple, so I turned to the harbour-master and asked him how often the south wind blew. Whereupon he replied that it was a very rare occurrence—about once a year, in fact. From this it looked as though we were going to be at La Luz for a long time. However, for the time we were going to forget our worries, and we determined to enjoy the pleasures of Las Palmas, and the great reception arranged for us by the Yacht Club.

The next day I was ill. I think it must have been caused by the culmination of over-exertion since we had left the Ivory Coast, so I lay in bed and endeavoured to clear up the accumulation of correspondence and detail by dictating the replies to my wife. However, being in bed had its advantages, for it saved me innumerable interviews, as my wife allowed me to deal with only the most urgent and important affairs.

Naturally my thoughts turned to the ' get-off ' question. With the impossibility of the north runway, and the impracticability of the east and west, where there was no runway at all, it was evident that our only chance was to the south, and as the south wind never blew I began to inquire for the next best thing in such circumstances—namely, a dead calm.

Although a flying-boat when heavily loaded cannot take off down wind it is possible for it to get off in a dead calm; therefore, if the north wind should cease to blow, our only hope, failing a south breeze, was a calm.

I consulted the meteorological expert on this matter, whereupon this gentleman replied that he would go away to his observatory in the mountains and consult his charts. He returned the next day with the good news that on the day after the morrow,

WORRALL INSPECTS THE TANK AT LA LUZ

at dawn, as the sun rose, the north wind would suddenly drop and cease to blow for a period of from eight to nine minutes, after which it would come up again in full force.

This was our chance, and we immediately set forth to make preparations. The harbour-master did everything possible to help us, and as we should need the entire length of the harbour he set to work and had 20,000-ton liners shifted in order to give us a straight fairway from the stone wall of the inner dock right to the mouth of the harbour.

Our fairway looked like a small version of a street in New York, for the liners towered like skyscrapers on either side of us, and it was going to be a difficult task to steer dead straight between them, as there was only a few feet to spare on either side of our wing-tips.

The day came. We were all up at four o'clock, and before daybreak were on board the Singapore, which had been tied up at the extreme end of the fairway overnight in readiness. There was no room to taxi about on the water to get our engines warm, therefore she had to be lashed astern to bollards on the quayside while our engines were started up.

One great main hemp-line, big enough to hold a liner, held the Singapore secure against the strain caused by our turning propellers, and it was arranged that at a given signal a man in a skiff under our tail should cut us adrift with a sharp knife.

It so happened that our fairway was parallel to the great long jetty that formed the harbour, and Worrall and I, on the previous day, had walked the whole length of this jetty, and had more or less measured our estimated take-off. We both realized the danger of not coming off the water and running into the open sea, where the big Atlantic rollers would certainly smash our hull.

We therefore agreed that we should have to fix on a point at which, if the flying-boat was not off the water, we could shut off our engines and come to a standstill before the open sea was reached. At the spot that we estimated we should have to be off the water or otherwise shut off our engines there happened to be a railway truck full of banana crates awaiting shipment on the jetty. This would be easily visible from the cockpit of our flying-boat, and we looked upon it as the safety mark.

We were all on board, and everything was stowed away, when daybreak appeared on the eastern horizon. We instantly started our engines, for sunrise quickly follows daybreak in this part of the world. The crew were all pessimistic about the wind question, for the north wind was still blowing, and there seemed no earthly reason why it should suddenly cease. Slowly our engines got warm, and I feared they would not be hot enough by the time the great moment arrived.

Presently the sun came up over the horizon, and like a miracle the wind dropped dead. Immediately the excitement on board was intense, and in order to hasten the warming process on my machine I opened the throttles as far as I dared, for to have overdone it would have meant running the risk of tearing off the tail of the machine, which was held fast by the hawser.

The engines were warm, and we were all ready to go, when we were attracted by the shouting of our friends from the quayside, who were gesticulating and pointing to the north. Instinctively I knew what they meant: the north wind was coming again, so without waiting another second I gave the signal by putting my hand above my head, whereupon the man on the tail cut the line, Worrall pushed the throttles open, and we bounded forward.

That take-off was one of the tensest moments of our whole flight round Africa. My feet were pressing hard on the rudder-bar, for the slightest swerve to the right or left would have brought our wing-tips into contact with the medley of shipping, and I think that the steering during the first one or two seconds was the most difficult, until we had some way on. Soon we were racing forward between the shipping and were up on the step of our hull, skimming over the water faster and faster, the mouth of the harbour getting bigger and bigger every moment.

Worrall kept saying, " She's got it. She's got it "; and as I could not see the railway truck with the banana crates out of the corner of my eye I imagined we had the affair well in hand. We tried to ease her off, but it was too soon and she would not have it. Then with a quick glance to my left I realized that some fool had moved the railway truck and it was too late to shut off, and our only hope was to get off before we finally reached the harbour mouth.

I held her lightly to the last fraction of a second until we were through the harbour mouth, when we hit the first swell, and I instantly pulled her off the water and she bounced into the air. There was a tense moment when everybody on board literally held their breath, waiting for the next thud should we not be off. Seconds seemed like minutes, but the dreaded bump never came. We were off, and, as my wife told me afterward, the long faces of all the crew broke into smiles and then jubilation.

Soon La Luz was fading away behind us, and we were heading eastward over the open sea past the southern end of the Island of Fuerteventura, toward the coast of Africa at Cape Jubi.

This trip from the Canary Islands back to the African coastline and up to Casablanca, although very peaceful from the flying point of view, was in reality a dangerous part of the journey, because, until we reached Mogador, just below Casa, there was no place whatsoever where we could have found sheltered waters from the Atlantic rollers to have landed our flying-boat. Engine-failure on this part of the journey might have meant the spoiling of our whole trip.

In due course we sighted Cape Jubi, and as we neared the mainland we could see that this desert coastline was relieved by just one thing, a French fort which had an aerodrome alongside. The aerodrome is one of the refuelling stations of the French Air Mail Service to Dakar, and as we flew low over this outpost all the garrison turned out, and by their

HOMING OVER THE ATLANTIC

gesticulations we guessed that they were sending us up a cheer.

Then came a desolate coastline, broken only by an inlet known as Puerto Cansado. The chief thing about the place was its name, and the fact that, had we landed there, surrounded by the limitless desert, we should have laid ourselves open to attack by extremely hostile Arab brigands.

The first sign of life and natural habitation that we met was at the little Spanish territory of Ifni, and here we espied a few Arab huts, for the desert was now changing from barren earth and yellow sand to occasional scrub and bush. Soon we were passing over a coastline of a sparsely populated area, and away inland I could see the walled-in Arab town of Tiznit, a place I had visited some eight years before when I had wandered through Morocco and Northern Africa with an aeroplane.

Then came the Bay of Agadir, that little port that became so famous in diplomatic circles throughout the world in the year 1911.

It was at Mogador that we encountered fog, the only place, with the exception of the south-west coast, where we had seen this weather condition on our whole journey round Africa. We tried to fly through it, but soon found ourselves enveloped completely, so we just opened our engines and climbed for all we were worth until we got above it. After passing for about ten miles with this white, woolly bed of vapour beneath us, we found the coastline all clear ahead once more, and thus came to the completion of that day's journey at Casablanca. I had been to Casablanca several times, and the town seemed to have doubled itself in size on each visit. On this occasion we had the pleasure of landing in the great new magnificent harbour.

Chapter XII

FAREWELL TO AFRICA

FROM this point onward we felt that we were back in civilization, and that our journey through the wilds was over. We enjoyed the wonderful hospitality of the British Consul; we were received by all the French dignitaries; in fact, our day's stay there was crowded with appointments.

The short stage from Casablanca to Gibraltar was of particular interest to me, for every landmark was so familiar. In November 1921, accompanied by one named Hatchett, we had inaugurated the first Spanish Civil Air Line between Seville and Larache, in Morocco, which service is still running to-day, Bill Hatchett having carried on with absolute regularity since that date.

Soon we reached Tangier, the north-west corner of the African continent. It lay beneath us, making a perfect aerial vista, with its whitewashed Moorish houses standing out brilliantly in the sunshine. Among the palm-trees and wooded hills were dotted, here and there, wonderful little palaces of the colony, which are there mainly because of the ideal climate.

As we passed over the Straits of Gibraltar we said good-bye to Africa, for ahead was Europe and a different world entirely.

In the past few months we had flown along the north coast of Africa, up the whole length of the Nile to its source, and then had jumped from lake to lake through Central Africa, until we came to the Indian Ocean. Then we had flown down the east coast, round the south coast, and then up the entire length of the west coast, having made for the first time in history a circumnavigation of the African continent in a flying-boat.

At Gibraltar the weather was fine, and on landing in the harbour we were not unduly troubled with the contrary winds for which this place is famous, although it is true that at one time we saw two flags, only a few hundred yards apart, both blowing toward one another, while the Singapore was constantly changing her position at her mooring-buoy.

I always remember that at Gibraltar my wife gave the canaries the first bit of greenstuff they had had since they left Cape Town. They, the gift of the King's Harbour - master at Malta, were destined to survive the entire journey; in fact, they are fit and strong to-day. The many visitors that went over our flying-boat at all the various ports concocted all sorts of stories about our canaries, and my crew overheard two dear ladies state with authority that the reason that we kept the canaries was in order that we should know when we were flying in bad air, for then the canaries would die, and thus we should know it was time to come down. I always look upon our birds as the world's champions, for they have flown farther than any other canaries in the world's history.

IN A FLYING-BOAT

For some contrary reason the sun did not shine on our journey up the east coast of Spain, but instead we had dull, overcast skies with a howling head wind for the best part of the trip. However, we reached Barcelona in safety, and on the following morning flew round the eastern end of the Pyrenees, a few miles from where Hannibal made his crossing some two thousand years before.

Once more we crossed the south of France and arrived at Bordeaux, the only place we visited twice on the whole trip. The officers and men at the French seaplane base at Hourtin gave us such a hearty reception, and once again we went back to the little inn where a few months before we had all dined together on our first night out from England at the start of our great venture.

Sheaves of telegrams had to be answered. We discovered that arrangements had been made for our arrival at Plymouth on the next day but one, and furthermore that we were expected to tour the coast-line of the British Isles so that the public might see Britain's first all-metal flying-boat after its 23,000-mile journey round the African continent.

The tour was to be from Plymouth to Southampton, then to Rochester, Hull, the Tyne, Leith, Glasgow, Belfast, Liverpool, Cardiff, Southampton, and finally back to Rochester, where we looked forward to landing on the Medway in front of Messrs Short Bros.' works, where the Singapore had been designed and built.

The programme had been arranged for us to arrive

at Plymouth by noon, so, wishing to be on time, we were up soon after dawn, making an early start, for one can always lose time in the air, whereas it might be more difficult to gain it.

All the crew were very bright on this particular day, especially Conway, whom we chided for having become engaged to the " only girl in the world," all by correspondence, during the flight. Bonnett, on the other hand, seemed to carry on with his daily duties just as though he were on the start of the flight; in fact, had it been necessary for us to have about turned and started all over again I feel quite sure that Bonnett would have betrayed no astonishment or emotion whatsoever. Nazner and Smith, who had joined us on the west coast, were now a little more cheerful, especially as their sunburnt knees had at last resumed their normal size, while Worrall and myself were happy at nearing home, although, in view of the fact that we had to go round the whole of the British Isles, we knew that our task was still far from completion.

I think that my wife had looked forward to this day so long that when it at last came she had no sensations on the subject whatsoever, and remarked that she would not consider that she was home until she had finally stepped off the machine at Rochester. For my own part my one ambition was to hand the machine over to the Air Ministry, for the Air Council, at the instigation of the then Secretary of State for Air, Sir Samuel Hoare, had approved of the scheme of lending me the Singapore, which was their property,

HOME AT LAST: DISEMBARKING AT PLYMOUTH

Sport and General

for this expedition round Africa. Now that the flight was practically accomplished I was confident that they would consider it had been worth while, for the reports and data that we had been able to collect of the performance of the aircraft and engines under the varying climatic conditions and other circumstances would be invaluable to the Air Ministry and the future of British flying-boat design; in fact, an asset to the whole aircraft industry.

As we neared Brest the overcast skies started to mingle with the mist off the sea and visibility became very bad. The cloudbanks seemed to mount up for thousands of feet, and so our only course was to go underneath them. But as soon as we turned the north-west corner of France we discovered that the Channel was covered in one of its noted fogs.

Before starting we had worked out our compass-bearing to take us direct to Plymouth, and, hoping that things might improve, we pushed out over the open sea. Instead of getting better things got worse, until we were flying only about ten feet off the waves into a thick driving fog, being unable to see more than a couple of hundred yards ahead of us, which is practically no distance whatever at ninety miles an hour.

The situation was getting serious, because if we accidentally collided with any of these big rollers underneath us, flying as we were across wind, we should have run the risk of serious damage, but what I most dreaded was that we should meet a ship or fishing-boat, for then we should have been on top of

247

it before we had time to turn. So, nodding to Worrall to keep his eye on the compass, I started on a flat turn in order to make my way back to the coast, where the fog was not quite so thick.

Our turn had to be a flat one, because had we banked the slightest bit, owing to the fact that we were only a few feet off the water, we should have put our wing-tip into the waves.

Presently Worrall signalled to me to steer straight ahead, as we were now flying in the reverse direction, and after a little while the visibility got slightly better. Soon we found ourselves skirting along the rocky coastline of Brittany, flying eastward. At times the visibility improved a little, only to get bad again shortly afterward, and eventually, realizing how treacherous the fog might be, I was determined to take refuge in the first sheltered water that we could find.

Just after passing Pointe du Château we could see ahead what appeared to be a blank wall of fog, and so, as there was a deep inlet at the sheltered mouth of the river Jaudy, we lost no time in turning head into wind and landing. After taxi-ing on the water for a little while we anchored midstream opposite a few cottages. My wife and I decided to go ashore in order to get a telegram off to Plymouth to tell them of our whereabouts and the cause of our delay, and soon discovered that this meant a journey of some dozen kilometres in a motor-car. However, we got to the village, and, despite the fact that the post-office was closed in the middle of the day, managed to get our telegram off. We then went into the nearest restaurant, ordered food and wine to take back to our crew, and while the basket was being prepared we had a light luncheon ourselves.

We could not help but think what an extraordinary circumstance it was that, after journeying for over 20,000 miles round Africa, we had never once been held up by weather. In fact, with the exception of a few hours' wait on a west-coast lagoon during a tropical storm, the only other delay for weather had been on the first day of the flight, when we were unable to leave the Southampton Water owing to the fog. We were now on the final stage, and the same Channel fog that had prevented us from starting on the first lap was now trying to stop us from completing the last lap.

By the time we got back to the Singapore a wind had suddenly sprung up from the north-east. It was, in fact, blowing a gale, with the result that the fog was lifting fast.

We scrambled on board, and, confident of improving conditions, started our engines up and taxied down the river, head into the north-easter. As the weather appeared to be far better away to the east we set out on a compass-course for Guernsey, and flew low through the driving rain and drizzle, and as soon as we sighted the island we headed north-west for the coast of Devon. Rain fell in torrents, but the visibility was good enough to see a ship half a mile ahead of us, which at that moment seemed a very friendly sight.

249

Suddenly the great black cloud mass ahead of us began to take shape, and soon we discovered that it was not a cloud at all, for through the heavy rain the cliffs of Devon were looming. It did not take us long to recognize Start Point, and very soon we were altering our course and racing by familiar landmarks, until suddenly the weather cleared and ahead was the Breakwater and Plymouth Sound. A few more minutes and we were circling over old Eddystone, and then came the landing in front of the Hoe.

Plymouth was the start of a sequence of magnificent welcomes and receptions, but unfortunately, from the moment we arrived in England, illness overcame each member of the crew. This may have been due to the sudden change of climate, for although it was the month of June our tour round the British Isles was accomplished in rain, fog, drizzle, and gale—conditions which were more in keeping with the depths of winter than the " month of roses "—until our final landing at Rochester, when the sun burst forth.

When at last the expedition was over we were happy in the thought that we had successfully accomplished a valuable flight of survey, investigation, and negotiation, and had laid the foundation of the great British air line, the Through-Africa Air Route.

Some of the other titles in the Adventure Travel Classic series
published by The Long Riders' Guild Press.
We are constantly adding to our collection, so for an
up-to-date list please visit our website:
www.thelongridersguild.com

The Rob Roy on the Jordan	John MacGregor
In the Forbidden Land	Henry Savage Landor
From Paris to New York by Land	Harry de Windt
My Life as an Explorer	Sven Hedin
Elephant Bill	Lt.-Col. J. H. Williams
Fifty Years below Zero	Charles Brower
Quest for the Lost City	Dana and Ginger Lamb
Enchanted Vagabonds	Dana Lamb
Seven League Boots	Richard Halliburton
The Flying Carpet	Richard Halliburton
New Worlds to Conquer	Richard Halliburton
The Glorious Adventure	Richard Halliburton
The Royal Road to Romance	Richard Halliburton
My Khyber Marriage	Morag Murray Abdullah
Khyber Caravan	Gordon Sinclair
Servant of Sahibs	Rassul Galwan
Beyond Khyber Pass	Lowell Thomas
True Stories of Modern Explorers	B. Webster Smith
Call to Adventure	Robert Spiers Benjamin
Heroes of Modern Adventure	T. C. Bridges
Death by Moonlight	Robert Henriques
To Lhasa in Disguise	William McGovern
The Lives of a Bengal Lancer	Francis Yeats-Brown
Twenty Thousand Miles in a Flying Boat	Sir Alan Cobham
The Secret of the Sahara: Kufara	Rosita Forbes
Forbidden Road: Kabul to Samarkand	Rosita Forbes
I Married Adventure	Osa Johnson
Grey Maiden	Arthur Howden Smith
Sufferings in Africa	Captain James Riley
Tex O'Reilly – Born to Raise Hell	Tex O'Reilly and Lowell Thomas

The Long Riders' Guild
The world's leading source of information regarding equestrian exploration!
www.thelongridersguild.com

CPSIA information can be obtained at www.ICGtesting.com
Printed in the USA
LVOW08s2321190514

386511LV00001B/84/P